CLIMATE CHANGE
THE FACTS 2017

CLIMATE
CHANGE
THE FACTS 2017

EDITED BY
JENNIFER MAROHASY

CONTRIBUTORS
John Abbot Sallie Baliunas Simon Breheny Bob Carter
Paul Driessen Tony Heller Craig Idso Clive James
Bjørn Lomborg Jennifer Marohasy Pat Michaels
John Nicol Jo Nova Ian Plimer Tom Quirk Peter Ridd
Matt Ridley Ken Ring Nicola Scafetta Willie Soon
Roy Spencer Jaco Vlok Anthony Watts

Institute of Public Affairs

Copyright © 2017
Institute of Public Affairs
All rights reserved.

ISBN: 978-0-909536-03-9

This edition published in 2017 by Connor Court Publishing Pty Ltd.

Institute of Public Affairs
Level 2, 410 Collins Street
Melbourne VIC 3000
Phone: 03 9600 4744

www.ipa.org.au

Connor Court Publishing Pty Ltd
Po Box 7257
Redland Bay QLD 4165

www.connorcourt.com

This book is dedicated to the memory of Bob Carter.

Contents

Foreword

Next year, 2018, the Institute of Public Affairs will celebrate its 75th year. The IPA was founded on the principles of freedom and a determination to, wherever possible, assist in enabling humans to flourish.

Our predecessors at the IPA were quick to see the dangers to freedom posed by the catastrophism and policy prescriptions emerging in the early days of what was then called global warming. A 1992 article in the *IPA Review*, our keystone publication, appropriately took a sceptical view, providing extensive references by way of contrast to the previous climate scare, which had been about a looming ice age. It was clear that something other than scientific advances were behind the abrupt shift from one kind of catastrophism to another.

The economic absurdity of the measures proposed to curtail emissions was also examined, and it was shown that even on the IPCC's premises the choice to engage in costly reductions would have only a marginal impact on any climate outcomes. In the intervening twenty-five years we are fortunate that Bjørn Lomborg has perfected that form of logical analysis, which proceeds by initially accepting the premises of the IPCC's climate forecasts, then demonstrating why the measures it proposes would have very large costs but only marginal impacts on climate. Twenty-five years on from the *IPA Review* article, we are thrilled his analysis of the Paris Accord appears as Chapter 15.

In subsequent years, it became apparent that not only was our freedom and our prosperity under threat, but that also the cherished traditions of the scientific method – which is itself a key inheritance of western civilisation – were facing collateral damage as the peddlers of alarmism began manipulating data, intimidating sceptical academics, and shutting down free speech in order to advance their agenda.

The IPA recognised that it was more important than ever for policy makers to have access to the facts and the latest research regarding climate change science and policy. The original *Climate Change: The Facts* was published in 2012, with another successful edition following in 2014. You are now reading the third book in the series. This edition, like its predecessors, will be sent to every member of the Australian Parliament to ensure they have at their disposal the facts regarding climate change.

This book would not have been possible without the singular efforts of Dr Jennifer Marohasy, who as editor has assembled a stellar list of authors. She has worked with them to drive this project to publication.

Jennifer has been associated with the IPA for more than a decade, publishing many works on environmental science addressing climate change and other shibboleths of environmental alarmism. The IPA is extremely grateful that Jennifer's return to the IPA in 2015 and the research she has since conducted was enabled by the philanthropic contribution of the The B Macfie Family Foundation, which also enabled the employment of Dr John Abbot and Dr Jaco Vlok (whose work appears in this volume). We would like to acknowledge and honour the continuing support of Bryant and Louise Macfie for all that is best about our heritage of free and honest scientific inquiry.

We also express our thanks to the more than 700 people who contributed to the research and related activities that enabled the publication of this book, through an appeal that was launched in late 2016. The success of this appeal enabled us to commit to a full-colour publication for the first time, allowing us to highlight and guide the reader through the rich data that underpins many of the key findings.

As well, we give thanks to the hardworking team members who have assisted the Editor in various stages of development and production, including Leonie Ryan (typesetting), Susan Prior (copy), Sandra Anastasi (graphics) and, at the IPA, Dr Chris Berg, Samy Therone, Josh Stranger, Darcy Allen, and Rachel Guy.

This work is very special to all of us at the IPA as it is dedicated to our esteemed former colleague Bob Carter. Bob was a brilliant scientist of the highest integrity, an excellent communicator, an indefatigable fighter for the truth, and – as we say in Australia – a top bloke. It is fitting that – with Anne Carter's permission – we are able to reproduce one of Bob's typically forthright, informative, and well-written works as Chapter 21.

We commend to you *Climate Change: The Facts 2017*.

Scott Hargreaves
Executive General Manager

John Roskam
Executive Director

Contributors

John W Abbot

John is a Senior Fellow at the Institute of Public Affairs with more than 100 peer-reviewed publications in international scientific journals – and more than a dozen recent publications in climate science. He is a graduate in chemistry from Imperial College (London), with a PhD from McGill University (Montreal). John is particularly interested in understanding kinetic and mechanistic behaviour associated with complex chemical and physical systems – most recently rainfall.

Sallie L Baliunas

Formerly a researcher at the Harvard–Smithsonian Center for Astrophysics, Sallie was the Deputy Director of the Mount Wilson Observatory from 1991 to 2003. In 2003, she coauthored a paper with Willie Soon in *Climate Research*, which concluded: 'The 20th century is probably not the warmest nor a uniquely extreme climatic period of the last millennium.' The paper's publication became controversial, forcing the resignation of five of the journal's editors – this was not due to any errors in the paper, which has never been retracted and remains an important contribution to climate science.

Sallie has a PhD in astronomy from Harvard University.

Simon Breheny

Director of Policy at the Institute of Public Affairs, Simon has a Bachelor of Arts and Bachelor of Laws from the University of Melbourne. While completing this study, he was elected President of the Melbourne University Law Students' Society, and appointed Vice-President of the Victorian Council of Law Students' Societies. He regularly appears on Australian television and radio as an advocate for freedom of speech. In 2016, he was elected Chairman of the International Young Democrat Union, the official youth organisation of the International Democrat Union.

Bob Carter

Bob was a climate scientist with particular expertise in stratigraphy, palaeontology and marine geology, and more than 40 years of professional experience. A graduate of Otago University (Dunedin) and Cambridge University (UK), he held tenured academic staff positions at Otago and James Cook University (Townsville), where he was Professor and Head of School of Earth Sciences between 1981 and 1999. He also served as chair of the Earth Sciences Discipline Panel of the Australian Research Council, chair of the National Marine Science and Technologies Committee, director of the Australian Office of the Ocean Drilling Program, and co-chief scientist on ODP Leg 181(Southwest Pacific Gateways).

Bob was emeritus fellow and science policy adviser at the Institute of Public Affairs, science adviser at the Heartland Institute, the Science and Public Policy Institute, and chief science adviser for the International Climate Science Coalition. Bob was the author of *Climate: The Counter Consensus* (2010) and coauthor of several more books, including *Taxing Air: Facts and Fallacies about Climate Change* (2013) and three volumes in the *Climate Change Reconsidered* series produced by the Nongovernmental International Panel on Climate Change (NIPCC).

Paul Driessen

Paul is a frequent guest on radio talk shows across the USA, and participates in energy, health and environmental conferences. His articles have appeared in many newspapers including the *Wall Street Journal, Washington Times, Investor's Business Daily, and New York Post.* His book *Eco-Imperialism: Green Power – Black Death,* documents the harm that restrictive environmental policies often have on poor families in developing countries, and has been printed in the United States and India and translated into Spanish, Italian and German.

Paul received his BA in geology and field ecology from Lawrence University and a JD from the University of Denver College of Law, with an emphasis on environmental and natural resources law.

Tony Heller

Often considered a heretic by the orthodoxy on both sides of the climate debate, Tony lives in Boulder (Colorado) where he rides his bicycle everywhere. He is a lifelong environmentalist – first testifying at a Congressional subcommittee hearing at age 15 in Kanab, Utah, in support of a wilderness area. Tony has degrees in geology and electrical engineering, and has worked as a contract software developer on climate and weather models for the United States government. Tony is perhaps best known as a blogger using the pen name 'Steve Goddard'.

Craig D Idso

Craig is the founder, former president and current chairman of the board of the Center for the Study of Carbon Dioxide and Global Change. He received his Bachelor of Science in geography from Arizona State University, his Masters of Science in agronomy from the University of Nebraska (Lincoln), and his PhD in geography from Arizona State University – with his doctoral thesis entitled *Amplitude and phase changes in the seasonal atmospheric CO_2 cycle in the Northern Hemisphere.*

Craig is the author or coauthor of several books, including *CO₂, Global Warming and Coral Reefs* (2009), *Enhanced or Impaired? Human Health in a CO2-Enriched Warmer World* (2003), and *The Specter of Species Extinction: Will Global Warming Decimate Earth's Biosphere?* (2003).

Clive James

Poet, author, critic, broadcaster, and translator – Clive James has always been a keen observer of popular culture and unsustainable fads. He has written more than forty books. His translation of Dante's *The Divine Comedy* and his 2015 collection *Sentenced to Life* were both *Sunday Times* top ten bestsellers. In 1992 he was made a Member of the Order of Australia and in 2003 he was awarded the Philip Hodgins Memorial Medal for literature. In 2012 he was appointed CBE and in 2013 an Officer of the Order of Australia. He holds honorary doctorates from Sydney University and the University of East Anglia.

Clive has previously written only incidentally on climate change.

Bjørn Lomborg

Adjunct Professor at the Copenhagen Business School and President of the Copenhagen Consensus Center, Bjørn is former director of the Danish government's Environmental Assessment Institute (EAI). He became internationally known for his best-selling book *The Skeptical Environmentalist* (2001). A year after its publication he founded the Copenhagen Consensus, a project-based conference where prominent economists establish priorities to improve conditions for the world's poorest people – most recently with a focus on Bangladesh and Haiti.

Jennifer Marohasy

Jennifer is a Senior Fellow at the Institute of Public Affairs, with recent publications in the international climate science journals *Atmospheric Research* and *Advances in Atmospheric Research*, as well as *Wetlands*

Ecology and Management, Human and Ecological Risk Assessment, Public Law Review and *Environmental Law and Management.* Jennifer has a Bachelor of Science and PhD from The University of Queensland (Brisbane) and a life-long interest in natural history and long-range weather forecasting.

Patrick J Michaels

Director of the Centre for the Study of Science at the Cato Institute (Washington DC), and author of *Shattered Consensus: The True State of Global Warming* and *Climate Coup: Global Warming's Invasion of Our Government and Our Lives,* Pat was a research professor of environmental sciences at the University of Virginia from 1980 to 2007.

John Nicol

John is a physicist, former dean of science at James Cook University (Townsville), and in retirement, chair of the Australian Climate Science Coalition. His PhD focused on electromagnetic theory. John went on to study the spectroscopy of gases, with a focus on spectral line broadening – including four years in the Clarendon Laboratory, at Oxford University (UK). Spectroscopy is an important, but little studied, component of the analysis of the effects of atmospheric carbon dioxide on climate.

Joanne Nova

Jo is perhaps best known as a prolific and popular blogger on science-related issues. Her blog won Best Topical Blog of 2015, the Life Time Achievement Award in the 2014 Bloggies and Best Australian and New Zealand Blog in 2012.

She has a Bachelor of Science from the University of Western Australia where she won the FH Faulding and the Swan Brewery prizes. A former associate lecturer of science communication at the Australian National University, Jo wrote *The Skeptic's Handbook* – translated into French,

German (twice), Swedish, Norwegian, Finnish, Turkish, Japanese, Danish, Czech, Portuguese, Italian, Balkan, Spanish, Lao and Thai.

Ian Plimer

Ian is Australia's best-known geologist. He is Emeritus Professor of Earth Sciences at the University of Melbourne, where he was Professor and Head of Earth Sciences after serving at the University of Newcastle as Professor and Head of Geology. He was Professor of Mining Geology at The University of Adelaide and in 1991 was also German Research Foundation research professor of ore deposits at the Ludwig Maximilians Universität, München. He was an editor for the five-volume Encyclopedia of Geology. He has written 10 books for the general public, including the international bestseller, *Heaven and Earth*.

Tom Quirk

Tom trained as a nuclear physicist at the University of Melbourne, has attended the Harvard Business School, and has been a fellow of three Oxford Colleges. During a long professional career Tom at various times worked for resources company, CRA (now known as Rio Tinto), in the United States at Fermilab, at the universities of Chicago and Harvard, and at CERN in Europe. He has held several positions in utilities associated with electricity generation, including a founding directorship of the Victorian Power Exchange.

Peter Ridd

Peter is Professor of Physics at James Cook University (Townsville) with particular interests in coastal oceanography, including human impacts on coral reefs, and instrument development for geophysical sensing. His consultancy work is associated with marine dredging operations, and the development of specialist instrumentation, with the profits used to fund student scholarships and research projects. More recently Peter has been promoting the requirement for better quality assurance in all areas of science.

He has published over 100 papers in international science journals.

Matt Ridley

British journalist, businessman, libertarian, Matt is the author of many popular science books including *The Red Queen: Sex and the Evolution of Human Nature* (1994), *Genome* (1999), *The Rational Optimist: How Prosperity Evolves* (2010) and *The Evolution of Everything: How Ideas Emerge* (2015), which have sold more than one million copies and been translated into more than 30 languages. Since 2013, Ridley has been a Conservative hereditary peer in the House of Lords.

Matt has a PhD from Magdalen College, Oxford (UK).

Ken Ring

Back in the 1970s, Ken was so interested in natural climate cycles, particularly lunar cycles, that the first word spoken by his infant son was 'Moon'. At that time, the family were living a subsistence existence on a remote stretch of the North Island of New Zealand. Ken learnt to read the weather with the help of the local Maori fishermen, who planned their activities according to the seasons – which they observed as being influenced by both the sun and the moon. When Ken eventually returned to 'civilisation', he went on to became a commercially successful long-range weather forecaster – publishing annual *Weather Almanac*'s for New Zealand, Ireland and Australia, and several books, including *The Lunar Code* (2006).

Nicola Scafetta

Currently Associate Professor in the Department of Earth Sciences, and at the Meteorological Observatory, University of Naples Federico II in Italy, Nicola worked as a research scientist in the physics departments at Duke University (Durham), and at the University of North Carolina (Chapel-Hill and Greensboro), Elon University (North Carolina), and in the ACRIM Lab (California). (ACRIM was NASA's first flight experiment dedicated to the task of monitoring the total solar irradiance output reaching the Earth.)

Nicola developed Diffusion Entropy Analysis, a method of statistical analysis that distinguishes between Levy Walk noises and Fractional Brownian motion in complex systems.

Wei-Hock (Willie) Soon

One of the world's best-known dissidents in climate science, Willie is an astronomer at the Mount Wilson Observatory, receiving editor with the journal *New Astronomy*, and researcher at the Solar, Stellar, and Planetary Sciences Division of the Harvard–Smithsonian Center for Astrophysics.

Willie coauthored the book *The Maunder Minimum and the Variable Sun–Earth Connection* and the textbook *Astronomy and Astrophysics*.

Roy Spencer

Before becoming a Principal Research Scientist at the University of Alabama in Huntsville in 2001, Roy was a senior scientist for Climate Studies at NASA's Marshall Space Flight Center, where he and John Christy received NASA's Exceptional Scientific Achievement Medal for their global temperature monitoring work with satellites. Roy is the United States' Science Team leader for the Advanced Microwave Scanning Radiometer flying on NASA's Aqua satellite. Roy received his PhD in meteorology at the University of Wisconsin–Madison in 1981.

Jaco Vlok

Jaco has degrees in electronic engineering from the University of Pretoria, South Africa, and a PhD from the University of Tasmania. He has undertaken research in radar and electronic warfare, including computer simulations, laboratory tests and field trials with South Africa's Council for Scientific and Industrial Research (CSIR). He is currently involved with a research program at the University of Tasmania exploring historical temperatures, with a particular focus on developing alternative techniques for historical temperature reconstructions using artificial neural networks.

Anthony Watts

Anthony is a certified television meteorologist, author of *Is the US Surface Temperature Record Reliable*, and founder of the world's most viewed climate website, *Watts Up with That*. Anthony has a life-long interest in weather stations, custom weather monitoring and recording, weather data processing systems, and weather graphics creation and display. He is also an advocate and user of renewable energy – particularly solar.

Introduction

Dr Jennifer Marohasy

There are certain things best not discussed with neighbours over the fence, at barbeques and at gatherings of the extended family; these topics used to include sex and politics, but more recently climate change has become a sensitive issue and has, consequently, crept onto the best-to-avoid list. At the same time as climate change has assumed this status, it has become a topic more likely to be included in a church sermon. Indeed, while once considered the concern of scientific institutions, climate change is now increasingly incorporated into faith-based initiatives with even Pope Francis weighing in, issuing an encyclical on the subject as explained in chapter 16 by Paul Driessen.

There are those who believe Pope Francis, and admire another climate change exponent, Al Gore – who marketed *An Inconvenient Truth* with comment, 'the fact of global warming is not in question' and that 'its consequences for the world we live in will be disastrous if left unchecked'. And then there are the die-hard sceptics who dare to doubt. Many claim that these climate sceptics and their support base have an undue political influence, successfully thwarting attempts to implement necessary public policy change.

This book is a collection of chapters by so-called climate sceptics. Each writer was asked to write on an aspect of the topic in which they are considered to have some expertise. *None* of them deny that climate change is real, but instead, they point out how extremely complex the topic of Earth's climate is, with some of the contributors also querying the, often generally accepted, solutions.

As you will see, this is not a book with just one message, except perhaps that there is a need for more scrutiny of the data, and of our own prejudices. This book's reason for being is to give pause for thought, and to throw some alternative ideas and considerations into the mix.

Natural climate cycles

Fundamental to the climate sceptics' perspective on climate change is the fact that there are natural climate cycles, and that even the most extreme portrayals of late-twentieth-century warming fall within what we might expect from a natural warming cycle. As one of Australia's best known geologists Ian Plimer explains in chapter 20:

> Climate change has taken place for thousands of millions of years. Climate change occurred before humans evolved on Earth. Any extraordinary claim, such as that humans cause climate change, must be supported by similarly extraordinary evidence, but this has not been done. It has not been shown that any measured modern climate change is any different from past climate changes. The rate of temperature change, sea-level rise, and biota turnover is no different from the past.
>
> In the past, climate has changed due to numerous processes, and these processes are still driving it. During the time that humans have been on Earth there has been no correlation between temperature change and human emissions of carbon dioxide (CO_2). Past global warmings have not been driven by an increase in atmospheric CO_2.
>
> Without correlation, there can be no causation.

Professor Plimer was a colleague of the late Bob Carter, to whom this book is dedicated.

Professor Carter was a director of the Ocean Drilling Program – an international cooperative effort to explore and study the history, composition and structure of the Earth's ocean basins. As Professor he had scrolls of data – time series – from the expeditions that he led as part of this program. I spent an evening with him at his home poring over one of his charts – a proxy record of New Zealand's climate over the past several thousand years, and we discussed how in this record from a sediment core the temperature could be seen to oscillate – there were natural climate cycles. His time series, or charts, for particular geographical locations, were printed out on long rolls of paper stretching the length of his kitchen table, and more. The end of the chart – that portion representing the present – was often dangling somewhere near the floor. These records provide an indication of the rate and magnitude of temperature change, as explained in chapter 21 – which is reprinted from his first book *Climate: The Counter Consensus*.

In an earlier chapter, Dr Nicola Scafetta, from the University of Napoli Federico II, discusses these natural cycles and explains how variations in solar luminosity – caused by the gravitational and electromagnetic oscillations of the heliosphere due to the revolution of the planets around the Sun, and even the tidal effects of the Moon – can drive oscillations in climate (chapter 3).

In chapter 4, Ken Ring, a commercially successful New Zealand based long-range weather forecaster, explains how the orbit of the Moon around the Earth can have a significant effect on local weather, with climate being the sum of all these weather events.

The physics of carbon dioxide

We are conditioned by the nightly news to believe that there is something extraordinary about current temperatures. There are any number of university professors – often quoted as part of an alleged 97% consensus on climate change – who assert a claimed catastrophic temperature increase from a doubling of atmospheric carbon dioxide (CO_2). Interestingly, those who speak publicly with most conviction – often quoting the

United Nations' Intergovernmental Panel on Climate Change (IPCC) – are from university geography departments. Yet it is an understanding of spectroscopy, normally the concern of analytical chemists and physicists, that is fundamental to explaining the likely impact of changes in the composition of the Earth's atmosphere on temperature.

Professor Svante Arrhenius was a chemist, who, in 1896, more than 120 years ago, was the very first person to propose that a doubling of atmospheric CO_2 could lead to a 5 to 6 °C increase in global temperatures. His calculations were speculative, and undertaken before modern high-resolution spectroscopy, which has enabled the measurement of the absorption and emission of infrared radiation by CO_2. This is explained by Dr John Abbot, an analytical chemist, and Dr John Nicol, a physicist with a background in spectroscopy, in chapter 19. Indeed, measurements from spectroscopy suggests that the sensitivity of the climate to increasing concentrations of CO_2 was grossly over-estimated by Professor Arrhenius, and these overestimations persist in the computer-simulation models that underpin the work of the IPCC to this day.

Yet it is the output from these same computer-simulation models – wanting a solid experimental foundation in radiative physics – which are used to generate criteria for studies in many other disciplines. For example, the IPCC General Circulation Models define the low, medium and high scenarios for the study of ocean acidification.

The Great Barrier Reef and ocean acidification

Ocean acidification is an area of research where, in less than 20 years, the number of published papers has increased from about zero each year to nearly 800 (Abbot and Marohasy, chapter 2). Ocean acidification is sometimes referred to as global warming's evil twin; and, of course, most of these 800 peer-reviewed papers published each year will emphasise the detrimental effects of an assumed reduction in pH – often based on output, yet again, from a computer simulation, extrapolating from laboratory experiments. In the case of ocean acidification, the scientists might have even added some hydrochloric acid to artificially reduce the pH of the water.

Science is currently funded and reported in such a way that inconvenient facts are more often ignored – and agreement with popular theory is emphasised.

Professor Peter Ridd, James Cook University, suggests this situation needs to stop if we are to address real and pressing problems, as opposed to wasting resources on invented issues (chapter 1). He makes this case with particular reference to the Great Barrier Reef, where he shows that not only are there the normal science distorting factors – such as only being able to get funding where there is a problem to be solved – but there is also the problem that many marine scientists are emotionally attached to their subject.

The economics of climate change

CO_2 is not only a so-called greenhouse gas, it is also a plant food. In chapter 13, Dr Craig Idso, from the Center for the Study of Carbon Dioxide and Global Change, emphasises the rather extraordinary productivity improvements over recent decades – particularly in agricultural crop yields – which he attributes in large measure to the powerful and positive effect of rising levels of atmospheric CO_2.

Dr Matt Ridley – with Bachelor of Arts and Doctor of Philosophy degrees from Oxford University, a science journalist, and a member of the United Kingdom's House of Lords – begins chapter 14 with a similar assessment: the claim that global warming is actually doing more *good* than harm – particularly through greening the planet.

In chapter 15, Dr Bjørn Lomborg of the Copenhagen Business School, explains the economics of the Paris Accord; he claims that adhering to the Accord is going to be very expensive while hardly affecting the global climate at all. Which is perhaps why so many are increasingly investing so much in attempting to close down open and honest debate.

There is no unifying theory of climate

Simon Breheny, from the Institute of Public Affairs, explains in chapter 17 that some in academia are leading the charge to extend the criminal law to the punishment of climate heresy on the basis of moral negligence. They advocate that the law should extend to all activities that seek to undermine the public's understanding of the 'scientific consensus'. Of course, the notion of a 'consensus' is fundamental to modern politics, but is generally alien to traditional science – at least Enlightenment science as practised by true sceptics.

In reality, and to paraphrase Dr Willie Soon, from the Harvard–Smithsonian Center for Astrophysics, and Dr Sallie Baliunas, formerly the deputy director of the Mount Wilson Observatory, writing in chapter 11, there is – as yet – no coherent theory of climate. Rather, as Dr Pat Michaels, a senior fellow at the Cato Institute explains in chapter 18, the mainstream climate-science community is wasting much time cherrypicking data from weather balloons and satellites, all in an attempt to make the temperature data consistent with a particular, and somewhat broken, paradigm that places too much emphasis on CO_2.

An advantage of my approach in the compiling of the chapters for this book – an approach where there has been *no* real attempt to put everything into neat boxes – is that there are many surprises. I am referring to the snippets of apparently anomalous information scattered through the chapters. These can, hopefully, one day, be reconciled. As this occurs, we may begin to see the emergence of a coherent theory of climate – where output from computer-simulation models bears some resemblance to real-world measurements that have not first been 'homogenised'.

There are many chapters in this book about 'homogenisation' (chapters 5, 6, 7, 8 and 9 by Anthony Watts, Tony Heller, Dr Tom Quirk, Jo Nova and me, respectively). Homogenisation, in essence, involves the remodelling of data, and is now a technique integral to the development of key official national and global measures of climate variability and change – including those endorsed by the IPCC.

It is generally stated that without homogenisation temperature series are unintelligible. But Dr Jaco Vlok from the University of Tasmania and I dispute this – clearly showing that there exists a very high degree of synchrony in all the maximum temperature series from the State of Victoria, Australia – beginning in January 1856 and ending in December 2016 (chapter 10). The individual temperature series move in unison suggesting they are an accurate recording of climate variability and change. But there is no long-term warming trend. There are, however, cycles of warming and cooling, with the warmest periods corresponding with times of drought.

Indeed, some climate sceptics consider the homogenisation technique used in the development of the official temperature trends to be intrinsically unscientific. They consider homogenisation a technique designed to generate output consistent with the computer-simulation models, which, in turn, are integral to the belief that there are consistent year-on-year temperature increases – contrary to the actual measurements. Temperature series that are a product of homogenisation could be considered 'alternative facts' – although, ironically, this is a term newly minted by those who generally agree with these self-same homogenised (remodelled) temperature constructs.

Conversely, many so-called sceptics will argue that the solution is to simply focus on the satellite data; however, this temperature record only begins in 1979. The satellite data is, nevertheless, a very valuable resource for understanding global and regional temperature change over the last nearly 40 years – as explained by Dr Roy Spencer, Science Team leader for the Advanced Microwave Scanning Radiometer on the National Aeronautics and Space Administration's (NASA) Aqua satellite, in chapter 12.

Conclusion

In the introduction to his newly published *Collected Poems*, Clive James – the author of so many phenomenal bestsellers in the UK and Australia – writes that at the end of a long life, and despite illness, he has kept writing for the last six years because, 'there were still some subjects waiting for their proper expression, so really I was beginning again'.

Clive James is a literary giant – his works are examined and re-examined regularly by the literary elite – but until now he has written only incidentally on climate change. In the final chapter of this book, James writes on exactly this subject. He acknowledges that he is no expert on computer-simulation modelling: 'I speak as one who knows nothing about the mathematics involved in modelling non-linear systems.' However, as he says: 'But I do know quite a lot about the mass media, and far too much about the abuse of language. So I feel qualified to advise against any triumphalist urge to compare the apparently imminent disintegration of the alarmist cause to the collapse of a house of cards.'

Clive James effectively places the current obsession with catastrophic climate change – and the imminent demise of the Great Barrier Reef – in a broader cultural context.

Which brings me back to the opening paragraph of this introduction – and to our conversations with neighbours, at barbeques, and at family gatherings. I cannot guarantee that after reading this book you will be better equipped to negotiate the politics of climate change. But hopefully you will have a better appreciation of the depth, breadth and complexity of the subject.

1 The Extraordinary Resilience of Great Barrier Reef Corals, and Problems with Policy Science

Professor Peter Ridd

The Great Barrier Reef is often used to show the imminent crisis we are supposedly facing from climate change. It is photogenic, the water sparkles blue, the fish and corals are beautiful and delicate, and most who see it – particularly marine biologists – fall in love with it. It is abhorrent to even contemplate that it could be destroyed or damaged by humanity.

The claimed imminent peril faced by the Great Barrier Reef has captured the public's imagination. When then US president Barack Obama visited Australia, he remarked that he wanted global action on climate change, so that maybe his daughters would have a chance to see the Great Barrier Reef. A visiting architect to my university revealed that his daughter, on discussing the latest reef bleaching event at school, came home depressed that she would probably never be able to see the Great Barrier Reef. Most of the world's population seems to have been persuaded that it has no more than a few years left.

There is no doubt that every decade or so, abnormally high seawater temperatures can cause corals to bleach (Marshall & Schuttenberg 2006). Bleaching is when the coral expels the symbiotic algae (zooxanthellae) which normally live inside an individual coral polyp. The polyps are the animals, generally a few millimetres across, that make the calcium carbonate structure of the coral.

Thousands or even millions of polyps make up an individual coral. The symbiotic algae live inside the polyp and make energy from sunlight; they share this energy with the polyp in exchange for a comfortable environment. However, when the water gets much hotter than normal, something goes wrong with the symbionts and they effectively become poisonous to the polyp. The polyp expels the symbionts and – because it is the symbionts that give the polyp its colour – the coral turns white. Without the symbionts, the polyp will run out of energy and die within a few weeks or months, unless it takes on more symbionts that float around naturally in the water surrounding the coral.

The ghastly white skeletons of bleached coral, particularly when seen on a massive scale, make graphic and compelling images to demonstrate the perils of climate change. The fact that this only happens when the water gets much hotter than normal makes it a plausible hypothesis that coral bleaching is caused by anthropogenic climate change. It is also often claimed by scientists that mass bleaching has only occurred since the 1970s, and that it is a recent phenomenon that did not occur 100 years ago when the water temperature of the Great Barrier Reef was 0.5 °C to 1.0 °C degrees cooler (Hughes 2016).

Despite this apparently plausible hypothesis, it will be argued in this chapter that there is perhaps no ecosystem on Earth better able to cope with rising temperatures than the Great Barrier Reef. Irrespective of one's views about the role of carbon dioxide (CO_2) in warming the climate, it is remarkable that the Great Barrier Reef has become the ecosystem, more than almost all others, that is used to illustrate and claim environmental disaster from the modest warming we have seen over the course of the last century.

Corals like it hot

Most species of coral that live on the Great Barrier Reef also live in much warmer water, closer to the Equator around Indonesia and

Thailand, where the water temperature reaches 29.0 °C. Coral growth rates are closely linked to temperature, as shown in Figure 1.1. The warmer water allows the coral to grow faster and more prolifically than it does on the Great Barrier Reef. The Great Barrier Reef has a temperature range from an average 25.0 °C in the south to an average 27.4 °C in the north (Lough & Barnes 2000). Coral growth rates increase with temperature to well above these averages. So, for example, in the southern Great Barrier Reef (25.0 °C) the corals are calcifying at half the rate of corals in Indonesia and Thailand, as shown in Figure 1.1. Therefore, it might be predicted that a modest increase in temperature, of a few degrees, would allow corals to grow faster on the Great Barrier Reef.

Figure 1.1 Calcification rate versus water temperature for *Porites* corals

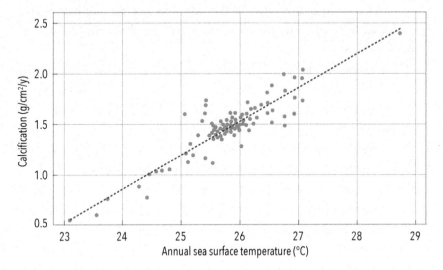

Source: Adapted by permission from Elsevier – Lough, JB & Barnes, DJ, 'Environmental controls on growth of the massive coral Porites', *Journal of Experimental Marine Biology and Ecology*, vol. 245, pp. 225-243, copyright 2000.

Corals are essentially a tropical species and the best corals live in the hottest places. Which is why there are some corals in Queensland that are regularly temperature stressed. In Moreton Bay near Brisbane, for example, they are stressed because the temperature gets too *low* in winter.

Juggling symbionts: incredible adaptability

Corals use a remarkable mechanism to reduce their susceptibility to large temperature changes (Steele 2016). There are a large variety of species of symbionts that can live inside particular species of coral. Some species of symbionts will allow the coral to grow faster, but will make them more susceptible to bleaching; other species of symbionts will give slower growth rates, but will make the corals relatively insensitive to extreme temperatures. Corals can select the species of symbionts that give them the ability to adapt to the prevailing conditions. However, it is always a gamble with the weather – if they choose the 'high octane' symbionts they could bleach; if they choose the safe 'low octane' symbionts and avoid bleaching they could be out-competed by neighbouring corals that grow faster.

The relationship between the polyp and the symbiont is central to the survival mechanism of corals, and is a masterpiece of adaptability. In their larval phase, most corals have no symbionts, but acquire them from the surrounding water, selecting the strain of symbionts to suit their conditions. In addition, a colony of coral may have a wide variety of symbionts in its multitude of different polyps. In the event of a severe bleaching, where much of the colony dies, those polyps with the strain of symbiont that mitigate against bleaching are able to regrow over the dead coral in the following years (Roff et al. 2014).

This ability to shuffle symbionts means that corals that undergo bleaching in one year will then be relatively unsusceptible to similar high temperatures in following years. The bleaching, in effect, forces the corals to take onboard a better adapted strain of symbionts.

The important point is that for a particular species of coral there is a variable upper temperature that they can tolerate. If these corals live in the generally cooler waters of the southern Great Barrier Reef, they may take onboard a species of symbiont that will mean they will bleach at 27.0 °C. This would be suicide in Thailand, but could be a good choice offshore from Gladstone, in the southern Great Barrier Reef.

It should be no surprise that corals have learned a thing or two about dealing with large temperature swings over 200 million years of evolution. But even if there was never any climatic variability, there is still a good reason for a particular coral to be able to deal with varying maximum temperatures. This is because coral spawn may drift large distances before they settle. It is therefore quite possible that the progeny of a particular coral could drift hundreds of kilometres (or further) into water that is hotter or cooler than where they originated. They could also drift into shallow water where the temperature is generally hotter, or they could settle in deeper, cooler, water. By simply varying the symbionts, young corals can deal with these different temperature regimes.

Great Barrier Reef corals truly are masters of temperature adaptability.

Bleaching is not a new phenomenon

Bleaching is one of corals' defence mechanisms and should be regarded as a strategy for survival rather than a death sentence. Generally, it stops them dying. Most corals that bleach fully recover (Marshall & Schuttenberg 2006) albeit they are a bit shaken by the experience. A survival mechanism such as bleaching indicates that corals have adapted to periods of unusually high temperatures in the past.

Professor Terry Hughes, a pre-eminent coral ecologist who works at James Cook University, Townsville, Australia, has claimed that bleaching is a new phenomenon. Professor Hughes has been responsible for much of the publicity about the 2016 bleaching event. He stated on Australia's ABC Radio National (RN):

A critical issue here is that these bleaching events are novel. When I was a PhD student 30 years ago, regional scale bleaching events were completely unheard of; they are a human invention due to global warming (ABC RN 2016).

In fact, bleaching was first recorded early last century by Sir Charles Maurice Yonge in the first major scientific study of the Great Barrier Reef (Yonge 1930). In addition, there are '26 records of coral bleaching before 1982' (Oliver et al. 2009). It was not until the 1960s that the phenomenon was discovered by scientists at the newly established institutions on the Great Barrier Reef coast – the Australian Institute of Marine Science, and James Cook University.

Spurious claims

Climate change and bleaching is only one of the latest threats to the Great Barrier Reef that scientists have been warning about. In the 1960s, it was claimed that the reef was being destroyed by plagues of crown-of-thorns starfish (COTS) (*Acanthaster planci*), in a similar fashion to plagues of locusts (Peason & Endean 1969). The cause of the plagues was immediately attributed to human activity, and a search for the specific culprit began. The first suspect was thought to be the overfishing of triton snails (*Charonia tritonis*), which eat COTS. More recently, it has been claimed, probably erroneously, that the run-off from agricultural land adjacent to the Great Barrier Reef, high in nutrients from the fertilisers used, is the cause (Brodie et al. 2007). In the meantime, we have learned that reefs rapidly recover from COTS outbreaks – within about ten years – and that the geological evidence suggests that COTS have been around for millennia – long before marine biologists first got hold of scuba gear (Walbran et al. 1989).

It is remarkable how rapidly some scientists have jumped to the conclusion that COTS outbreaks are a recent phenomenon.

It is noteworthy that there are strong parallels between scientists' reactions to COTS and to coral bleaching. If a mass bleaching event or a COTS outbreak had occurred on the Great Barrier Reef in the 1930s, who would have noticed? It is possible that a few pearl or *bêche-de-mer* divers might have, but they would have been unlikely to report their findings to scientists or the world's media. Contrast that to today, when a bleaching or COTS event will be documented by hundreds of scientists who may have been waiting and preparing for the event for years. They monitor the reef with satellite images of water temperature, giving them time to prepare massive aerial surveys of the reef. I've seen this research executed with military precision.

Spawning is not a new phenomenon, either

Bleaching is not the only visually spectacular, but recently discovered, event that occurs on the Great Barrier Reef. The other is the mass spawning event that occurs late in the year after a full moon. Almost all the corals on the Great Barrier Reef spawn on one or two nights of the year, making huge floating pink–white 'slicks' of eggs and sperm along the entire 2000 km length of the reef. From the air these slicks are seen as pink–white lines, often kilometres long and tens of metres across. From a boat, the slicks cannot be missed. However, this spawning event was not 'discovered' to science until the 1980s (Harrison et al. 1984).

So, did corals invent sexual reproduction in the early 1980s? Or is it more likely that it took a little while for scientists to discover something that many Aboriginal people, fishermen and other mariners, must have witnessed repeatedly in the past? The same applies to bleaching – it is a phenomenon only recently documented – not a phenomenon that has only recently occurred. It is interesting to note the different reactions of scientists to these two discoveries. Mass coral spawning is a wonder of nature and a result of millions of years of evolution; mass coral bleaching is new and caused by burning coal.

Corals: born together and die together

One of the features of coral reefs is the almost continuous change that occurs due to the succession of extreme events, of which high water temperature is just one. A picture-postcard reef today may be obliterated tomorrow by a large cyclone, a plague of COTS, a plume of freshwater from a nearby river during a flood event, or even from a period of cold water. The corals often all die together in a spectacularly massive event. However, in the last 40 years we have learned that they are also capable of rapid recovery – in a decade, or so.

It is also notable that the corals species that tend to bleach are the naturally short-lived species – generally, the plate and staghorn corals (Marshall & Schuttenberg 2006). The 'massive' corals, which resemble solid blocks or spheres and which can live for centuries, rarely bleach. The plate and staghorn corals are very susceptible to other events, such as cyclones, while the massive corals – which grow more slowly because they must lay down far more calcium carbonate in their skeleton – are not easily smashed by the waves from a cyclone. To grow perhaps 0.5 m above the sea floor, a massive coral will lay down between 10 to 100 times as much skeleton by mass.

The staghorn and plate-like corals have the shared philosophy of living fast and dying young. Pretty as they are, in many regards they are the weeds of the reef and we must not get too emotional when they get damaged by bleaching. The next cyclone will shortly spell their demise in any case. The longer lived massive corals are more akin to the giant trees in temperate forests that live hundreds of years. These corals are rarely bleached or killed by cyclones.

In between such destructive events, the reef quietly grows and waits for the beginning of the next cycle of death and regrowth. And, just recently, the attention of the world media fed by our science organisations.

Corals and ocean acidification

Rising water temperature is not the only way that CO_2 is predicted to kill the coral reefs. Much has been written about the effect of CO_2 on lowering the water pH, which it is claimed will retard the ability of corals to calcify or lay down their skeletons (De'ath et al. 2009). Ocean water is slightly alkaline (pH a little over 8, neutral is 7), and rising CO_2 concentrations will possibly drop this to a little under 8.

Changes in pH have already been claimed to have caused a calamitous change in coral calcification rates on the Great Barrier Reef – a drop of 15% from 1990 to 2005 (De'ath et al. 2009). Such claims, like so much research that supposedly shows a massive decline of the Great Barrier Reef, were received by the world's media with much fanfare. However, it is perhaps yet another example of science that has not been properly scrutinised, or subjected to proper quality assurance.

Like trees, which produce rings as they grow, corals set down a clearly identifiable layer of calcium carbonate skeleton each year, as they grow. The thicknesses and density of the layers can be used to infer calcification rates and are, effectively, a measure of the growth rate. Dr Glenn De'ath and colleagues from the Australian Institute of Marine Science used cores from more than 300 corals, some of which were hundreds of years old, to measure the changes in calcification during the last few hundred years (De'ath et al. 2009). They claimed there was a precipitous decline in calcification since 1990, as shown in Figure 1.2.

However, I have two issues with their analysis. I published my concerns, and an alternative analysis, in the journal *Marine Geology* (Ridd et al. 2013). First, there were instrumental errors with the measurements of the coral layers. This was especially the case for the last layer at the surface of the coral, which was often measured as being much smaller than the reality. This forced an apparent drop in the average calcification for the corals that were collected in the early 2000s – falsely implying a

Figure 1.2 Coral calcification rates

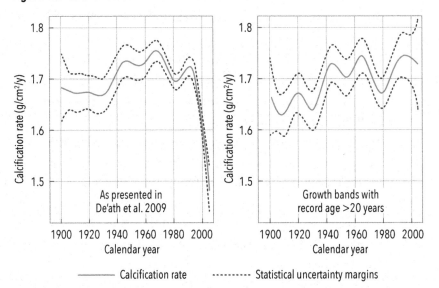

Coral calcification rates (left chart) as presented by De'ath et al. (2009), and (right chart) corrected for errors (Ridd et al. 2013). The dotted lines denote statistical uncertainty margins.

Source: Reprinted by permission from the American Association for the Advancement of Science – De'ath, G, Lough, JM & Fabricius, KE, 'Declining coral calcification on the Great Barrier Reef', *Science*, vol. 323, pp. 116-119, copyright 2009; and reprinted by permission from Elsevier – Ridd, PV, DaSilva, ET & Stieglitz, TC, 'Have coral calcification rates slowed in the last twenty years?' *Marine Geology*, vol. 346, pp. 392-399, copyright 2013.

recent calcification drop. Second, an 'age effect' was not acknowledged. When these two errors are accounted for, the drop in calcification rates disappear, as shown in Figure 1.2.

The problem with the 'age effect', mentioned above, arose because in the study De'ath and colleagues included data from corals sampled during two distinct periods and with a different focus; I will refer to these as two campaigns. The first campaign occurred mostly in the 1980s and focused on very large coral specimens, sometimes many metres across. The second campaign occurred in the early 2000s due to the increased interest in the effects of CO_2. However, presumably due to cost cutting measures, instead of focusing on the original huge coral colonies, the second campaign measured smaller colonies, many just a few tens of centimetres in diameter. In summary, the first campaign focused on large old corals, while, in contrast, the second campaign focused on small young corals. The two datasets were then spliced together, and wholly unjustifiable assumptions were implicitly made, but not stated – in particular that there is no age effect on coral growth (Ridd et al. 2013, De'ath et al. 2013).

Reporting good news as bad news

Dr Juan D'Olivo Cordero from the University of Western Australia collected an entirely different dataset of coral cores from the Great Barrier Reef (D'Olivo et al. 2013) to determine calcification rates. This study determined that there has been a 10% increase in calcification rates since the 1940s for offshore and mid-shelf reefs, which is the location of about 99% of all the coral on the Great Barrier Reef. However, these researchers also measured a 5% decline in calcification rates of inshore corals – the approximately 1% of corals that live very close to the coast. Overall, there was an increase for most of the Great Barrier Reef, and a decrease for a small fraction of the Great Barrier Reef.

While it would seem reasonable to conclude that the results of the study by D'Olivo et al. (2013) would be reported as good news for the Great Barrier Reef, their article in the journal *Coral Reefs* concluded:

> Our new findings nevertheless continue to raise concerns, with the inner-shelf reefs continuing to show long-term declines in calcification consistent with increased disturbance from land-based effects. In contrast, the more 'pristine' mid- and outer-shelf reefs appear to be undergoing a transition from increasing to decreasing rates of calcification, possibly reflecting the effects of CO_2-driven climate change.

Imaginatively, this shift from 'increasing' to 'decreasing' seems to be based on an insignificant fall in the calcification rate in some of the mid-shelf reefs in the last two years of the 65-year dataset. Why did the authors concentrate on this when their data shows that the reef is growing about 10% faster than it did in the 1940s?

Science quality assurance

In this chapter, I was asked to focus on climate change. I have highlighted just a few examples of questionable science – the list is long. Furthermore, climate change is only one of many claimed stressors causing damage to the Great Barrier Reef; others include sediments, nutrients and pesticides from agriculture. I have investigated these supposed threats and they are even less convincing and more contrived than the claimed effects of climate change (Ridd et al. 2011; and 2012). But challenges to the conventional wisdom are typically ignored, largely drowned out and sidelined by the majority. There is now an industry that employs thousands of people whose job it is to 'save the Great Barrier Reef'. As a scientist, to question the proposition that the reef is damaged is a potentially career-ending move (Lloyd 2016).

So, what is the solution?

The fundamental problem is that we can no longer rely on 'the science', or for that matter our major scientific institutions. There are major quality assurance shortcomings in the way we conduct what I will call 'policy science' – that is science used to inform public policy. In fact, in most cases, the only quality assurance measure is peer review. Peer review sounds impressive; I suspect the public thinks it is where, like a jury, a dozen scientists consider the scientific arguments and the data for many days before passing their verdict on whether it is good science or not. Unfortunately, peer review usually consists of a cursory read of the scientific paper, often for just a couple of hours, by two scientists. They never have the time to check the data properly, or to try to repeat the analyses. Their main task is to make sure that the writing and diagrams are clear, and that there are no obvious problems. Usually, we do not even know who these reviewers are. Is this the quality assurance process that we need if we are going to spend public funds on decisions that are supposed to be based on solid science?

In contrast to government policy science, research with an industry or medical focus usually includes some proper quality assurance, with good reason. For example, a company hoping to develop a drug from promising university trials will typically need a billion dollars to take it to market. The first step for the company is to check and replicate the original peer-reviewed research. It is of concern that when these checks are done, conclusions from the original work are found to be in error more than half the time (Prinz et al. 2011). This could be disastrous, but at least the checks were made to prevent wasting vast resources.

Policy science concerning the Great Barrier Reef is almost never checked. Over the next few years, Australian governments will spend more than a billion dollars on the Great Barrier Reef; the costs to industry could far exceed this. Yet the keystone research papers have not been subject to proper scrutiny. Instead, there is a total reliance on the demonstrably inadequate peer-review process.

The lack of quality assurance in science has become a hot topic, particularly in medical science. The failure of drug companies to replicate the findings of scientific institutions is just the tip of the iceberg. In the biomedical sciences, many authors have reported the level of irreproducibility at around 50% (Vasilevsky et al. 2013; Hartshorne & Schachner 2012; and Glasziou 2008). More recently, John Ioannidis, Professor of Medicine and of Health Research and Policy at Stanford University School of Medicine, and a Professor of Statistics at Stanford University School of Humanities and Sciences, suggested that as much as 85% of science resources are wasted due to false or exaggerated findings in the literature (Ioannidis 2014). Professor Ioannidis focused on, among other matters, the lack of funding for replication studies, which are so import-ant in the medical area. Indeed, replication of already 'known' results is one of the fundamental processes upon which the reliability of science rests, but this is generally seen as mundane and not the way to advance a scientific career. Funding bodies are rarely keen to spend money on such work.

The problem is so acute that the editor of *The Lancet*, one of medi-cine's most important journals, stated that:

> The case against science is straightforward: much of the scientific literature, perhaps half, may simply be untrue. Afflicted by studies with small sample sizes, tiny effects, invalid exploratory analyses, and flagrant conflicts of interest, together with an obsession for pursuing fashionable trends of dubious impor-tance, science has taken a turn towards darkness. (Horton 2015)

Similar concerns have also been raised for the psychological sciences (Wagenmakers et al. 2011).

How long will it take before we finally address this issue when it comes to policy science in general, and for the Great Barrier Reef, in particu-lar? Marine biologists Dr Mariana Duarte from the Federal University of Minas Gerais, Brazil, and Dr Howard Browman from the Institute of Marine Research, Norway, have called for 'organised scepticism' to improve the reliability of the environmental marine sciences (Duarte et al. 2015; and Browman 2016). Duarte et al. (2015) argue that:

> the scientific community concerned with problems in the marine ecosystem [should] undertake a rigorous and systematic audit of ocean calamities, with the aim of assessing their generality, severity, and immediacy. Such an audit of ocean calamities would involve a large contingent of scientists coordi-nated by a global program set to assess ocean health.

This is what must occur for the Great Barrier Reef. I have carried out half-a-dozen audits on some of the science claiming damage to the

Great Barrier Reef, and in every case I have discovered serious problems (Ridd 2007; Ridd et al. 2011, 2012, 2013). However, individuals can be easily ignored. There is a need for a properly funded group of scientists whose sole job is to find fault in the science upon which we are basing expensive public policy decisions regarding the Great Barrier Reef.

Conclusion

Due to the remarkable mechanisms that corals have developed to adapt to changing temperatures, especially the ability to swap symbionts, corals are perhaps the least endangered of any ecosystem to future climate change – natural or man-made. The corals found on the Great Barrier Reef also live in waters closer to the Equator, which are considerably warmer. Coral generally grows faster in warmer waters, so it should not be surprising that there has been a 10% increase in calcification rates at the Great Barrier Reef since the 1940s.

Yet, so many are convinced that the Great Barrier Reef is under threat due to the fact that when corals die, they tend to do it in spectacular ways with events that make excellent images for the media. Then there are the many 'scientific studies' that have never been replicated or properly checked, that conclude the imminent demise of the Great Barrier Reef.

There are serious problems with quality assurance in many areas of science, and possibly more so for Great Barrier Reef policy science. Not only are there the normal science distorting factors, such as only being able to get funding when there is a problem to be solved, there is also the problem that many marine scientists are emotionally attached to their subject. The world needs people who care for the environment; many of these scientists have signed up for a career of relative poverty to pursue marine biology. However, given these emotional pressures, together with the lack of a formal quality assurance mechanism, and documented examples of misinterpretation of calcification rates, we can be sceptical of claims that the Great Barrier Reef is in peril.

2 Ocean Acidification: Not Yet a Catastrophe for the Great Barrier Reef

Dr John Abbot & Dr Jennifer Marohasy

There has been an exponential increase in research on the topic of ocean acidification, which broadly concerns chemical changes in the ocean in response to increased concentrations of atmospheric carbon dioxide (CO_2). Uptake of CO_2 by the oceans from the atmosphere can potentially alter the balance of inorganic chemicals present, in turn affecting biological processes – including photosynthesis and calcification rates at coral reefs (Gattuso & Hansson 2011; Raisman & Murphy 2013). Coral reefs are a major focus of ocean acidification research. There is concern that increased atmospheric CO_2 will significantly reduce calcification and so negatively impact the overall health of these iconic ecosystems – held together by calcium carbonate that is secreted by corals.

In this chapter, we provide some background into the physical and chemical processes associated with ocean acidification, before considering the research into the effects of ocean acidification on biological organisms, with a particular focus on Australia's Great Barrier Reef.

Ocean acidification has been described as an impending 'ocean calamity', and the 'evil twin of global warming'. Public interest has fed the explosion of research on ocean acidification, considered unprecedented in the marine sciences (Browman 2016). Indeed, as shown in Figure 2.1, there were nearly 4000 articles published on ocean acidification between

Figure 2.1 Number of peer-reviewed papers published on ocean acidification

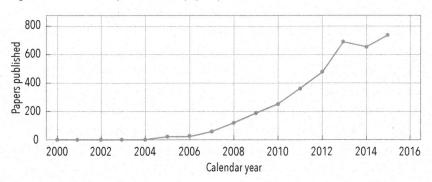

Source: Web of Science, which is an online subscription-based scientific citation indexing service, maintained by Thomson Reuters.

2000 and 2015. Most of the articles describe the effects of changes of pH on biological organisms; many of the claims are based exclusively on laboratory experiments (Riebesell & Gattuso 2015). However, a problem with laboratory experiments is that they cannot capture the complexities of the real world, not even the tremendous natural variability in ocean pH – which is a measure of ocean acidification.

Some basic chemistry

CO_2 occurs naturally in the Earth's atmosphere in very low concentrations, presently at about 400 ppm (parts per million).[1] Accurate measurements of atmospheric CO_2 concentration have been carried out at an observatory in Hawaii since 1959, when the concentration was 312 ppm. Pre-industrial levels are estimated to be about 280 ppm.

There is a natural exchange of gases between the atmosphere and dissolution in the ocean, with flows in both directions. CO_2 dissolves in water with its solubility depending on the temperature: the lower

---

1   Atmospheric Carbon Dioxide Record from Mauna Loa, https://climate.nasa.gov/vital-signs/carbon-dioxide/

the water temperature, the higher the solubility of a gas such as CO_2. An equilibrium may eventually be established when the flows in each direction are in balance. In turn, the strength of the flow in each direction is directly proportional to the concentration. If the concentration of atmospheric CO_2 increases, this increases the rate of CO_2 flow into the ocean, and eventually a new equilibrium establishes.

After dissolving in the uppermost layer of the ocean, CO_2 generates a number of changes to the chemical composition of the seawater. The chemistry of inorganic carbon in seawater is relatively complex, and can be understood in terms of a set of inter-related equilibrium processes. Dissolved inorganic carbon is mainly present in three chemical forms:

- aqueous (CO_2)
- carbonate ions (CO_3^{2-})
- bicarbonate ions (HCO_3^-).

There is also a minor amount present as carbonic acid (H_2CO_3).

So, carbon entering the ocean as CO_2 can be transformed into these other inorganic ions and molecules. When these reactions occur, hydrogen ions (H^+) are liberated into the seawater, and this in turn affects the pH of the seawater (pH is a measurement of the acidity of an aqueous solution, defined as the negative logarithm of the hydrogen ion concentration in the solution).

The term 'ocean acidification' refers to the decrease in pH that occurs when the concentration of CO_2 in the atmosphere increases through a chain of reactions that can cause the ocean pH to drop. A neutral solution (neither acidic nor alkaline) has a pH of 7.0. Acidic solutions have a pH less than 7.0, while alkaline solutions have a pH greater than 7.0. Typically, present-day seawater varies widely in its pH, but generally it is alkaline, having an average pH in the range of 8.2 to 8.4.

Ocean acidification does not imply a scenario where the oceans become acidic (that is, the pH of the ocean falls below 7.0). The whole process could also be referred to, in perhaps less emotive terms, as 'neutralisation', or alternatively 'carbonation' (Gattuso & Hansson 2011) because there is an increase in the concentration of dissolved inorganic carbon.

Another very important component of the inorganic system in seawater is calcium, which can exist as ions in solution (Ca^{2+}) and as solid calcium carbonate ($CaCO_3$). Calcification is controlled by the concentrations of calcium and carbonate ions in the seawater at a particular location. How much the seawater is saturated by these is known as the calcium carbonate saturation state, and it is described by the Greek term Ω (omega). The value of Ω depends on the particular natural form of calcium carbonate considered, commonly either aragonite or calcite. When Ω falls below 1, solid calcium carbonate dissolves.

Some concerns relate to the possibility of the saturation state decreasing as the pH falls. The key issue is, that as the concentration of carbonate ions declines, thereby reducing the value of Ω, it will affect the ability of organisms to undertake calcification, which could potentially have a very detrimental impact on shells and coral reefs.

Different processes, operating on very different timescales, become important in determining the fate of CO_2 in the atmosphere. Once released, it takes about a year for CO_2 to mix throughout the atmosphere. The ocean consists, broadly, of three layers: the uppermost, or boundary layer; the mixed layer, affected by waves and turbulence caused from wind stress on the sea surface; and the deep-ocean layer. The very uppermost boundary layer of the ocean almost instantaneously equilibrates with the overlying atmosphere. Below that, is the mixed layer, which then absorbs the CO_2 through the action of wind, waves and resulting turbulence. Once in the mixed layer, CO_2 can be transported to the deep ocean, through processes that occur on millennial timescales.

Variation in pH

Popular articles written for non-scientists often claim that an increase in atmospheric concentrations of CO_2 have already caused a decrease in oceanic pH of 0.1. These same articles claim that the Earth's oceans have a pH of 8.2 and that it will fall to pH 7.8 by 2100.

However, these are average numbers, which fail to provide any indication of the extent of natural variability. For example, early review articles of ocean chemistry, which predate current concerns with acidification, report pH levels of 9.4 in isolated coral reef pools during the warmth of the day, falling to 7.5 at night (Revelle & Fairbridge 1957). At night, organisms continue to respire CO_2, while there is no uptake of it through the mechanism of photosynthesis, which occurs during the day, hence the lower pH level. A recent study at Heron Island on the Great Barrier Reef suggested that pH generally falls below 8.0 soon after midnight in summer, while climbing to about 8.4 in the afternoon in winter, as shown in Figure 2.2 (Kline et al. 2015). Indeed, the night-time pH minima on the reef flat fringing Heron Island are already lower than the pH values predicted for the open ocean by 2100.

Figure 2.2 Variation in daily pH

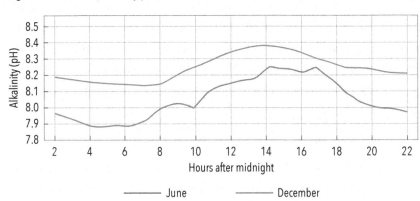

Source: Adapted with permission from Kline, DI et al., 'Six month *in situ* high-resolution carbonate chemistry and temperature study on a coral reef flat reveals asynchronous pH and temperature anomalies,' *PLoS ONE*, vol. 10, no. 6, e0127648, copyright 2015.

Direct continuous measurements of oceanic pH at the same location extend back only a few decades. For example, monthly data from the Hawaii Ocean Time-Series (HOT) programme only commenced in 1988. When values from this programme are averaged, the data indicates a downward trend in pH. This is in accordance with ocean acidification theory, as shown in Figure 2.3. But longer series based on proxy measurements suggest that this type of trend may simply be part of a natural cycle of rising and falling oceanic pH.

In order to understand changes over hundreds of years, approximate pH values have been estimated using boron isotopes in coral; these are known as proxy measurements. Data from such a study of proxies at Flinders Reef on the Great Barrier Reef shows that there have been periods of rising and also falling pH in the range pH 7.9 to 8.2 during the last three centuries – as shown in Figure 2.4 (Pelejero et al. 2005).

Figure 2.3 Monthly pH measurements

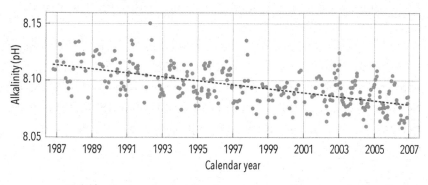

Measurements of pH for surface layers down to 30 m.

Source: Adapted by permission from PNAS – Dore, JE et al., 'Physical and biogeochemical modulation of ocean acidification in the central North Pacific', *PNAS*, vol. 106, no. 30, pp. 12235-12240, copyright 2009.

Figure 2.4 Changes in pH over decadal time scales

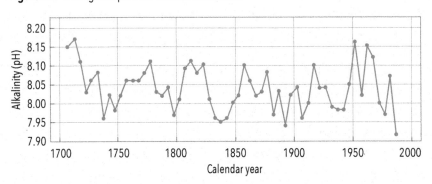

Variations in pH at Flinders Reef reconstructed from boron isotopes in coral.

Source: Adapted by permission from The American Association for the Advancement of Science –
Pelejero, C et al., 'Preindustrial to modern interdecadal variability in coral reef pH', Science, vol. 309,
no. 5744, pp. 2204-2207, copyright 2005.

These values correspond to the range of 7.9 to 8.3, which was reported from a similar study in the South China Sea with specimens that date back 7000 years (Liu et al. 2009).

These relatively long series indicate that comparatively rapid changes in pH have occurred on a decadal time scale before the recent increase in atmospheric concentrations of CO_2. For example, in Figure 2.4, consider the periods 1710 to 1750, 1815 to 1845, and 1950 to 1990. It is evident that the rate of decline in pH was approximately 0.05 pH units per decade in each case. This is more than the rate of decline recorded over recent decades at Hawaii of 0.033 pH units per decade, as shown in Figure 2.3.

It would therefore seem reasonable to conclude that there is nothing particularly unusual about the rate of pH decline over recent decades, because similar rates of change on a decadal scale occurred in the pre-industrial era.

Assessing claims of imminent catastrophe

The Great Barrier Reef is in the Coral Sea, off the coast of Queensland. It is the world's largest coral reef ecosystem comprising more than 2900 individual reefs and 900 islands stretching for 2300 km. The Great Barrier Reef is best known for its diversity of corals, but this ecosystem is also home to seagrass beds, which support significant populations of dugongs.

Review articles in popular scientific journals with titles like 'Ocean acidification hits Great Barrier Reef' have stated that coral growth has been sluggish since 1990 due to an increase in both sea temperatures and 'acidity' as a result of human-caused global warming (Biello 2009). Such claims are based on studies that in turn claim, unequivocally, that the recent decline in coral growth rates is unprecedented (De'ath et al. 2009). During the 2016 federal election campaign in Australia, university researchers quoted such studies as evidence that if there was not an immediate increase in funding the reef was doomed.

But there is reason to doubt the veracity of these claims.

Indeed, concerns about the lack of evidence underpinning many of the claims of imminent catastrophe due to ocean acidification have resulted in calls for 'organised scepticism' from within this same community (Browman 2016). The *Journal of Marine Science* recently published a special issue on the topic of ocean acidification with an introductory article by Dr Howard Browman from the Institute of Marine Research, in Norway. His article documents a long list of issues that need to be addressed if we are to have a reasonable level of confidence in the conclusions from ocean acidification studies.

Key issues that need to be considered when assessing claims of imminent catastrophe can be summarised under the following six headings:

1. Publication bias

Most published scientific studies on ocean acidification report negative effects of CO_2 on organisms, concluding that ocean acidification will be detrimental to marine ecosystems. Some of these studies, particularly those published in 'high impact' journals and featured in the mainstream media, predict an acidification-generated calamity in our oceans. As is true across all of science, however, studies that report no effect of ocean acidification are typically much more difficult to get published and, if published at all, seem to appear in lower-ranking journals. These are subsequently ignored by the mainstream media.

Studies in other related areas of science, for example in fisheries research, have shown that even when original studies have been proven incorrect, they often continue to be quoted in the media (Banobi et al. 2011).

2. Relevant timescales and exposure levels

Many early studies on ocean acidification exposed organisms to water chemistry that greatly exceeded even worst-case climate change scenarios. Not only have pH exposure levels been unrealistic, but studies continue to be undertaken over very short time periods, with inferences subsequently made about effects over decades or centuries. Such studies neglect the potential for acclimation, adaptation, or evolution.

Detailed studies on an important unicellular marine algae that produces calcite scales, *Emiliania huxleyi*, found that the immediate negative physiological response to ocean acidification could be partially compensated by evolutionary adaptation (Lohbeck et al. 2012; Lohbeck et al. 2014).

3. From single organisms to ecosystems

Most ocean acidification research is focused on a single species with investigations into their short-term physiological response. Understanding a whole-of-ecosystem response, however, often requires some understanding of the relative impacts on different species, including through competitive and trophic (feeding) interactions. Even if species are not directly affected by changing water chemistry, they may none-theless experience changes in abundance or distribution because their prey, predators, or competitors are affected. Although a study of shifts in the composition of corals in response to elevated CO_2 indicated that it was the individual tolerance of different species, rather than competition between species, that determined composition (Brien et al. 2016).

4. The need to consider variability in CO_2 and pH

It is well established that CO_2 and pH vary on a daily, seasonal, and inter-annual basis, including with fluctuations in temperatures, across bodies of ocean, and also with different depths (Hofmann et al. 2011; Waldbusser & Salisbury 2014). Yet experiments on biological systems have generally failed to incorporate this variability into the design of experiments, or into the interpretations of results (Eriander et al. 2016). Some researchers have pointed out that organisms exposed to large ranges in CO_2 and pH during their daily lives, life cycles and distribu-tional ranges, should be more tolerant of ocean acidification (Lewis et al. 2013).

The importance of a naturally fluctuating ocean pH is beginning to be recognised; for example, experiments on barnacles living in an environment with a fluctuating pH indicate environmentally and evolu-tionarily important responses (Eriander et al. 2016). However, very few studies have reported experimental investigations into the effects of simulated diurnal fluctuations in pH on organisms. Many of the studies on ocean acidification test the response of organisms to different fixed levels of CO_2 above the water in which the experiment is undertaken. The concentration levels of CO_2 are typically assigned:

*   ambient (400 – 450 ppm)
*   moderate (550 – 600 ppm)
*   high (900 – 1200 ppm).

5. Extrapolating from the laboratory

The Free-Ocean CO_2 Enrichment (FOCE) experimental approach has been developed to attempt to address some of the limitations of experiments undertaken in laboratories, by conducting ocean acidification experiments *in situ* – and by enabling the precise control of CO2 enrichment in partially open, experimental enclosures (Gattuso et al. 2014). The first FOCE systems were developed at the Monterey Bay Aquarium Research Institute, but others have since been deployed or are being planned around the world. Although straightforward in concept, the engineering and logistical aspects of FOCE technology are very challenging to implement. The key elements of FOCE experimental units include:

- partially open enclosures that allow for control of seawater conditions, but retain through-flow of ambient seawater
- a CO_2 mixing system
- sensors to monitor pH, as well as other critical environmental parameters
- a control loop to regulate the addition of gases or liquids to each experimental enclosure.

Such experiments on *Porites* corals from Heron Island reef flats show that this important species exerts strong physiological controls on the pH of their calcifying fluid (Georgiou et al. 2015). Over a six-month period, from mid-winter to early summer, the corals maintained their calcifying fluid pH at near-constant elevated levels, independent of the highly variable temperatures and FOCE-controlled carbonate chemistries to which they were exposed. This is an important finding because it implies that corals may have a high degree of tolerance to ocean acidification.

6. Methodological issues

Major methodological issues have been identified not just for some but for the vast majority of ocean acidification studies (Cornwall &

Hurd 2016). Problems included interdependent or non-randomly interspersed treatment replicates, and insufficient methodological detail.

The perception that calcification rates of corals at the Great Barrier Reef are in terminal decline due to ocean acidification is based on a study of 328 colonies of *Porites* corals from 69 reefs (De'ath et al. 2009). This study appears comprehensive and has been extensively reported in the mainstream media. However, after critical examination, the study has been shown to include basic flaws – including a systematic data bias in the last growth band of each core. As a consequence, the study erroneously concluded a recent drop in calcification rates (Ridd et al. 2013).

The potential impact will depend on the organism

Initial concerns about ocean acidification focused on organisms that construct their shells or skeletons from calcium carbonate. Such organisms are referred to as marine calcifiers and include not only corals, but also crabs, clams and conchs (sea snails).

Theoretically, and according to popular science magazines, all corals are already severely and negatively affected by ocean acidification. But this is not evident from methodologically sound studies undertaken at the Great Barrier Reef. A review of the growth rates of six, hard coral species at Lord Howe Island (Anderson et al. 2015) found marked variation in the growth rates of branching coral, while growth rates of the massive *Porites* coral were unchanged. The researchers suggested that a decline in the growth rates of the branching species could be attributable to a reduction in the calcium carbonate saturation state as a consequence of higher summer temperatures. A study measuring calcification rates for 41 long-lived *Porites* corals from seven reefs from the central Great Barrier Reef (D'Olivio et al. 2009), showed good recovery from the major 1998 bleaching event, with no significant trend in calcification rates for the inner reefs. Corals from the mid-shelf central Great Barrier Reef, however, did show a decline of 3.3%.

While most ocean acidification research has been focused on physiological processes, in particular calcification, there have also been studies on three common hard corals to look at their fertilisation, embryonic development, larval survivorship, and metamorphosis (Chua et al. 2013a; Chua et al. 2013b). These studies have found the early life-history stages were unaffected by reduced pH; there was no consistent effect of elevated CO_2 alone, nor in combination with temperature.

Studies of the effect of very high CO_2 levels (up to 2,850 ppm) on molluscs – including oysters, clams, scallops and conchs – have shown that these species will generally build their shells more slowly as CO_2 levels increase (Ries et al. 2009). This same study showed that crabs and lobsters respond quite differently to the same elevated CO_2 levels, showing a general increase in calcification rates.

The varied responses among different organisms reflect their differing abilities to regulate pH at the site of calcification, and:

- the extent to which their outer shell layer is protected by an organic covering
- the solubility of their shell, or skeletal mineral
- the extent to which they use photosynthesis (Ries et al. 2009).

Of course, many marine organisms are not calcifiers, and some of these organisms have also been tested for a response to ocean acidification.

When seagrasses collected from three locations in the Great Barrier Reef region – Cockle Bay, Magnetic Island, and Green Island – were exposed to four different CO_2 concentration levels for two weeks – with water temperature and salinity in the experimental tanks near-constant throughout – all three seagrass species exhibited enhanced photosynthetic responses (Ow et al. 2015). That is growth rates, observed after two weeks of exposure to an enriched CO_2 environment in an indoor aquarium, were higher. This suggests that ocean acidification could mean more seagrass, which would be good for large marine mammals like dugongs (dugongs are vulnerable to extinction because of issues unrelated to changing ocean chemistry).

Also, contrary to expectations, laboratory investigations into the effects of three different CO_2 treatments on anemonefish (commonly known as the clownfish) found that higher CO_2 levels stimulated breeding activity (Miller et al. 2013). The breeding pairs from the fringing reefs of Orpheus Island on the Great Barrier Reef, where they are exposed to the highest CO_2 levels, produced double the number of clutches per breeding pair, and 67% more eggs per clutch than the control. However, young anemonefish that were bred in high CO_2 levels and high temperatures showed decreases in their length, weight, condition, and survival (Miller et al. 2012). Though these effects were absent or reversed when their parents also experienced the higher concentrations (Miller et al. 2013).

Acidification in context

The concept of ocean acidification, and human-caused global warming more generally, has been described as containing a grain of truth embedded in a mountain of nonsense (Lawson 2008; Carter 2010). Indeed, the projected large increase in atmospheric CO_2 will at most cause a small reduction in pH – it will not turn the ocean acidic. Yet this is what is implied by the term ocean acidification. True acidification would require average pH to be reduced below 7.0, at which point shells would indeed begin to dissolve. This is an impossible scenario, however, because of the ocean's effectively limitless buffering capacity

This review focused on (what have been termed) ocean acidification studies on biological organisms, with particular reference to the Great Barrier Reef. It has highlighted the limitations of the current research and the difficulties associated with making generalisations. Most studies have been on single species in contrived laboratory conditions. They have been of short duration, and they have not considered the potential for adaptation. In the few instances where adaptation has been considered, it has been shown to significantly modify the impact of varying pH as a consequence of elevated levels of CO_2.

All of this needs to be assessed against the reality that along the length and breadth of the Great Barrier Reef there are naturally occurring large daily fluctuations in pH, and that it is unclear as to what extent the current trends of apparent pH decline are part of existing natural cycles.

3 Understanding Climate Change in Terms of Natural Variability

Dr Nicola Scafetta

Recent scientific research has shown that the Intergovernmental Panel on Climate Change (IPCC) climate models fail to properly reconstruct the natural variability of the climate at multiple time scales: a fact that is fundamental to properly interpreting climate changes. If the IPCC's models cannot reproduce the historical temperature patterns and trends correctly, then there is no reason to believe that they could reliably predict future temperature changes.

Advanced techniques of pattern recognition have shown that it is likely that natural climate variability consists of cycles. The most relevant ones can be described as oscillations spanning periods of 9.1, 10.5, 20, 60, 115, 900–1000 and 2100–2500 years. These oscillations are not only related to variations in solar luminosity and tidal effects of the Sun and of the Moon, but are also driven by gravitational and electro-magnetic oscillations of the heliosphere due to the revolution of the planets around the Sun (Scafetta 2013, 2014; Kirby 2007; Svensmark et al. 2009, 2012; Ollila 2015; Scafetta et al. 2016). In fact, in addition to luminosity and tidal forcings, interplanetary electromagnetic forcing is relevant because it can modulate cosmic ray flux and cosmic dust concentrations falling on Earth, which are likely responsible for Earth's cloud-albedo system modulation.[1]

Questioning the general circulation models

Analytic modelling of the climate, for example that undertaken by the Coupled Model Intercomparison Project Phase 5 General Circulation Models (CMIP5 GCMs), relies on a complex set of coupled differential equations describing the circulations of the atmosphere and of the ocean. And on a set of external radiative forcing functions influencing the system (IPCC 2013). If the forcing functions were constant, the climatic system would, at most, fluctuate around a mean value similar to a correlated red noise.[2] In order to obtain climate change within the models some time evolution of forcing function is required.

The typical forcing functions used in the climate models are those deduced from the changes of atmospheric greenhouse-gas (GHG) concentrations, such as carbon dioxide (CO_2) and methane (CH_4), as well as atmospheric aerosol concentrations, volcanic aerosol concentrations, land-use changes, total solar irradiance change, and other factors. The claim is that, with the exception of the solar and volcanic forcings, all the other listed radiative forcings have an anthropogenic origin, inasmuch as they originate from human activity. Note that water vapour (H_2O), the most important of the GHGs existing in the atmosphere, is not included among the models' forcings. This is because it is assumed to be a feedback of the system, the evolution of which is directly calculated by the thermodynamic equations of the models.

---

1   The proportion of solar energy reflected back into space is called the albedo; the presence of clouds increases the Earth's albedo.

2   Red noise is a type of signal noise characterised by long-range autocorrelations produced by fractal Brownian motion, hence its alternative name of random walk noise; red noise is stronger in longer wavelengths similar to the red end of the visible spectrum.

Figure 3.1 Comparing estimates of climate sensitivity

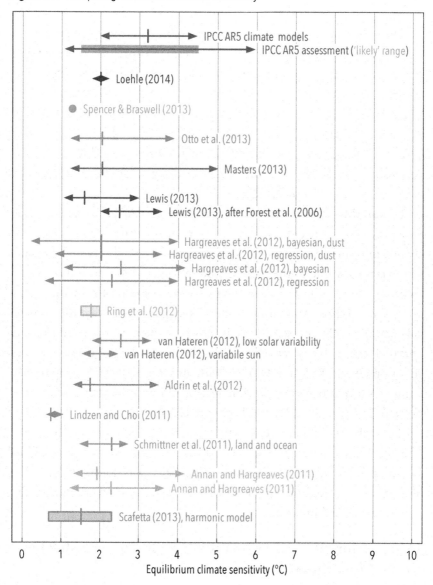

A comparison of the estimates of climate sensitivity to the radiative forcing induced by a doubling of atmospheric CO_2 concentration.

Source: Adapted from IPCC 2013, Scafetta 2013 and Lewis 2013.

The most important climate parameter to be determined is the climate sensitivity to the radiative forcing. This value measures the total climatic effects of the radiative forcings by taking into account the additional contribution from the internal feedbacks of the system (for example, water vapour, cloud cover, snow cover, and so on). The analytic climate models, such as the CMIP5 GCMs, estimate that if the atmospheric concentration of CO_2 doubles, the climate system would warm, at the equilibrium, by about 3 °C, with an uncertainty range spanning between 1.5 and 4.5 °C, as shown in Figure 3.1.

It should be stressed, however, that the error bar for this estimate is very large. The same radiative forcing could cause warming ranging from a given minimum value to a value three times larger. This range of uncertainty is mostly due to the models' poor ability to reproduce the various feedbacks, including the modelling of water vapour and of the cloud system. In addition, numerous studies have pointed out that the models' calculation of climate sensitivity is also very likely to be overestimated and that a more likely range for it could be about half of the above one, that is, between 0.75 and 2.25 °C, with an average of 1.5 °C. Figure 3.1 compares a number of these climate sensitivity estimates (IPCC 2013; Scafetta 2013; Lewis 2013; and other references therein).

Thus, the real climate sensitivity to radiative forcing could be about half of that estimated by the models that are trusted by the IPCC to interpret the twentieth century's warming. This casts doubt on their reliability, and on the objectiveness of their scientific predictions and the consequent social implications.

Evidence for Holocene climatic variability unreproducible by the climate models

It is relatively simple to demonstrate that the traditional analytic models are unreliable because they significantly overestimate the climate sensitivity to radiative forcings. This is shown by their inability to reconstruct past climate changes. In short, they fail to reproduce observed climatic variations. For example, Shakun et al. (2012) show that global warming was preceded by increasing CO_2 concentrations during the last deglaciation. Moreover, temperature reconstructions of the Holocene[3] suggest that after the Holocene maximum, which was between 7000 and 9000 years ago, the global surface temperature cooled by about 0.5 °C to 1.0 °C (Shakun et al. 2012; Marcott et al. 2013). In contrast, the climate models have erroneously shown a persistent warming trend of about 0.5 °C to 1 °C during the last 11,000 years, ostensibly because during this period an increase in the atmospheric concentrations of CO_2 has occurred (Liu et al. 2014). Figure 3.2 stresses the divergence between the temperature cooling trend (from the Greenland Ice Sheet Project 2 (GISP2) record, Alley 2004) and the increasing CO_2 record during at least 5000 to 6000 years before the current age, suggesting that CO_2 cannot be the main driver of climate change.

In fact, the Holocene has been characterised by large quasi-millennial temperature oscillations that do not correlate with the CO_2 record, as shown through a comparison of the two charts in Figure 3.2. The models also fail to reconstruct the large oscillation that has been responsible for the well-known Roman, Medieval (from the ninth to the fourteenth century) and Current (since the nineteenth century) Warm Periods; and the Dark Age (from the fifth to the eighth century) and Little Ice Age (from the late fourteenth to the eighteenth century) that were associated with cold periods (Ljungqvist 2010). The large millennial climatic oscillation is far better correlated with solar activity and cosmic ray flux variation indexes, as estimated using carbon-14 and beryllium-10 isotope records (Bond et al. 2001; Kerr 2001; Steinhilber et al. 2012; Kirkby 2007) than with CO_2 or other GHG records. Also, the 2100- to 2500-year Hallstatt oscillation – found in both the solar and in the

---

3   The Holocene is the last geological interglacial period that began about 11,700 years ago.

Figure 3.2 Comparing Holocene temperature and CO_2 records

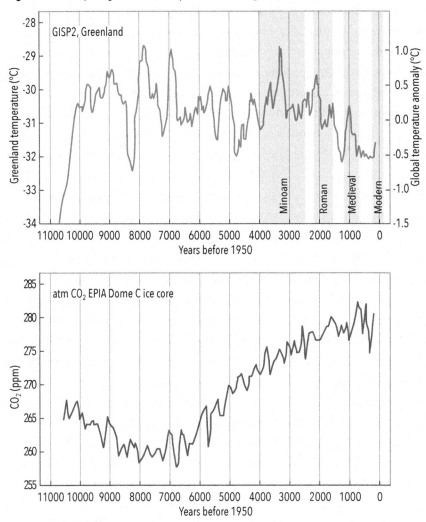

A comparison between the Holocene temperature record from Greenland (top) and the CO_2 record for the same period (bottom chart).

Source: Alley, RB 2004 (top chart); and bottom chart is adapted by permission from Liu, Z, et al., 'The Holocene temperature conundrum', *PNAS*, vol. 111, no. 34 copyright 2014.

climate proxy records throughout the Holocene – has recently been found to correlate to specific stable orbital resonances of the solar system (Scafetta et al. 2016).

The misleading hockey stick temperature graph

The inability of GCMs to reconstruct climate variability through the Holocene has been fully recognised only recently. Before the 1990s an approximate understanding of climate change during the last millennia had suggested a large pre-industrial climate variability (Lamb 1965; 1982). However, the inability of the climate models to reconstruct such variability has now become a big issue in climate science because, at least since 2000, the previous understanding of climate variability through this period has been replaced by an alternative one based on a global temperature reconstruction known as the 'hockey stick' graph (Mann et al. 1999; Crowley et al. 2000). This infamous chart is the top chart in Figure 3.3, and it only begins in 1000 AD.

The hockey stick graph depicts the global surface temperature remaining approximately constant from 1000 AD to 1900 AD, and then rising abruptly from 1900 (Mann et al. 1999; Crowley et al. 2000). This specific hockey stick pattern was well reconstructed by the first energy balance models attempting to interpret climate change (for example, Crowley et al. 2000), and it is still well reconstructed by the current anthropogenic global warming theory models, such as the CMIP5 models. In fact, these models predict that the natural climate variability is relatively insignificant relative to the warming from anthropogenic forcing (IPCC 2013). Therefore, the hockey stick temperature graph gives the illusion that the available climate models were sufficiently reliable in interpreting that the twentieth-century warming is due, almost entirely, to the increase of atmospheric GHGs.

However, since 2005, alternative reconstructions of Northern Hemisphere temperatures (more consistent with Lamb 1965; 1982) have been proposed (Moberg et al. 2005; Mann et al. 2008; Ljungqvist 2010; Christiansen & Ljungqvist 2012). These reconstructions demonstrate that the climate is characterised by a large millennial oscillation during the last 2000 years, as shown in Figure 3.3 (bottom chart) and also Figure 3.4. The natural climate variability manifested in these records is three to four times larger than that shown in the hockey stick temperature reconstructions. Therefore, the most recent global surface temperature reconstructions stress the importance of modelling the natural climate variability to correctly interpret climate change.

It should be noted that, while the hockey stick temperature reconstruction predicts a cooling of about 0.2 °C, or less, from the Medieval Warm Period to the Little Ice Age of the sixteenth to nineteenth centuries, the most recent reconstructions show three to four times larger variability of about 0.6 to 0.8 °C, in several regions of the Earth (Ljungqvist 2010).

The large climatic variability observed since medieval times can be correctly interpreted only if the climatic effects of solar variability are appropriately incorporated into the models. This effect has been severely underestimated in the climate models by three to six times and, simultaneously, the climatic effect of the radiative forcings, including the CO_2 forcing, has been overestimated by at least two times in the same models (Scafetta 2013).

In fact, the warming observed since 1900 started in the eighteenth century, that is, since the end of the Little Ice Age. Therefore, the climate began warming well before anthropogenic GHG emissions could have had any significant impact on the climate. Indeed, the modern warming period is very likely the result of a natural warming trend due to the quasi-millennial oscillation evident in solar and climate records throughout the Holocene (for example, Bond et al. 2001; Kerr 2001; Hoyt & Schatten 1997; Scafetta 2014) that should reach a maximum in the 21st century (Scafetta 2012d).

Figure 3.3 Comparing temperature records, previous 2000 years

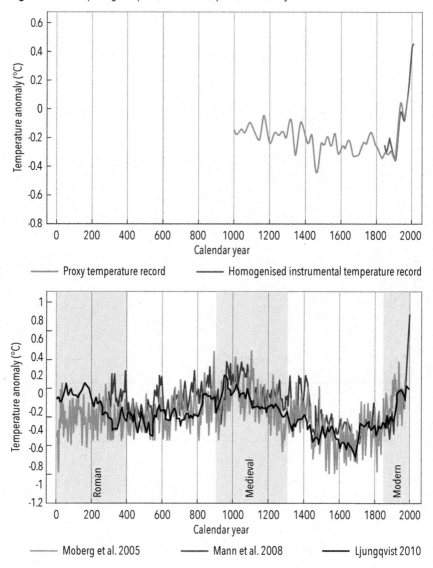

A comparison between the 'hockey stick' temperature reconstruction (top) and other global and Northern hemisphere temperature reconstructions (bottom). Note the 'hockey stick' only begins in 1000 AD.

Source: Top chart based on Mann et al. 1999 and Crowley et al. 2000. Bottom chart adapted with permission from Moberg, et al. 2005, 'Highly variable Northern Hemisphere temperatures reconstructed from low and high resolution proxy data', *Nature*, vol. 433, pp. 613-617, copyright 2005; and Mann, ME, et al., 'Proxy-based reconstructions of hemispheric and global surface temperature variations over the past two millennia', *PNAS*, vol. 105, no. 36, pp. 13252-13257, copyright 2008; and Ljungqvist, FC, 'A new reconstruction of temperature variability in the extra-tropical Northern Hemisphere during the last two millennia', *Geografiska Annaler*, vol. 92, pp. 339-351, copyright 2010.

Figure 3.4 Comparing temperature records, previous 1200 years

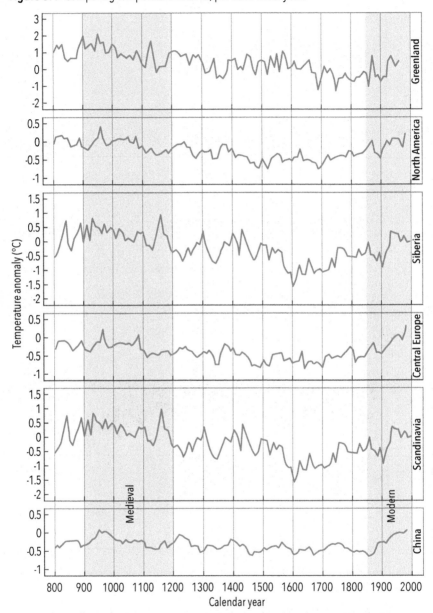

Regional temperature reconstructions from the Northern Hemisphere also show fluctuations, and a different general temperature profile when compared to the 'hockey stick'.

Source: Adapted with permission from Ljungqvist, FC, 'A new reconstruction of temperature variability in the extra-tropical Northern Hemisphere during the last two millennia', Geografiska Annaler, vol. 92, pp. 339-351, copyright 2010.

The decadal and multidecadal natural climate oscillations

Analysis of global surface temperature records – which have been recorded since 1850 (Morice et al. 2012) – suggests an overall warming trend of about 0.9 °C from 1850 to 2015, as shown in Figure 3.5. There are fluctuations within this record. Sets of major oscillations – with periods of 9.1, 10.4, 20, and 60 years – have been found to characterise this record (Scafetta 2010; 2012; 2013), with this natural variability

Figure 3.5 Official global surface temperature reconstructions

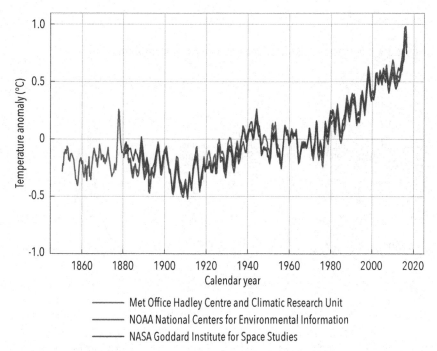

The 12-month homogenised moving-averages from the instrumental surface temperature record are shown for the period January 1850 to February 2017.

Source: https://data.giss.nasa.gov/gistemp/, http://www.metoffice.gov.uk/hadobs/hadcrut4/ and https://www.ncdc.noaa.gov/cdo-web/.

approximately harmonic, like the rhythmic variability of the ocean tidal motion, but on a much longer time scale. These rhythms could have an astronomical origin because similar harmonics are found among the main gravitational oscillations of the solar system (Scafetta 2010; 2012; 2013; 2014; 2015; 2016). The strong peak in 2015-2016 was due to the large El-Niño event.

My spectral analysis of the global surface temperature record, and also those of the Northern and Southern Hemispheres, are shown in Figure 3.6 (Scafetta 2010; 2013).

Figure 3.6 Spectral analysis of the global surface temperature records

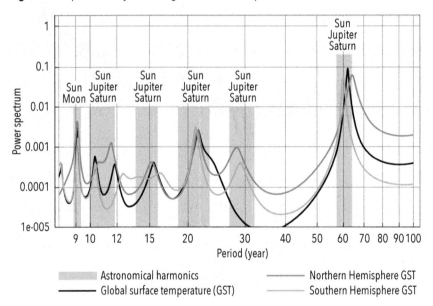

The orange area refers to known astronomical oscillations of the solar system induced by the Sun, the Moon, Jupiter and Saturn.

Source: Adapted from Scafetta 2013.

Spectral analysis can be used to find the major frequency components – the peaks of the power spectrum curve – of any signal. The fact that the Northern and Southern Hemispheres, which are otherwise quite different, show the same spectral peaks suggests that the two hemispheres are strongly coupled, and, as well, suggests that these oscillations have an astronomical origin.

A 60-year oscillation has been observed in a large number of climate indexes, including in the following:

- global and regional temperature records
- the Atlantic Multidecadal Oscillation (AMO) index
- the Pacific Decadal Oscillation (PDO) index
- the North Atlantic Oscillation (NAO) index
- Indian monsoon records
- and in many other indexes (Scafetta 2010, 2012b; Manzi et al. 2012; Wyatt & Curry 2014; Gervais 2016).

None of the general circulation models used by the IPCC has been able to properly model these oscillations that are evident in the surface temperature reconstructions (Scafetta 2012a; 2013).

The quasi-60-year cycle dominates the temperature variation. The 60-year oscillation was in its warm phase during the periods:

- 1850–1880
- 1910–1940
- 1970–2000.

And in its cooling phase during the periods

- 1880–1910
- 1940–1970
- and, also, it is likely since 2000.

This 60-year oscillation explains why the global surface temperature – the so-called temperature *hiatus* (Meehl et al. 2011) – has remained nearly stationary since 2000. The simple explanation is that since 2000, the 60-year oscillation, being in its cooling phase, has compensated for the anthropogenic warming that occurred during this same period. But, on the contrary, the GCMs have on average predicted a 2 °C per century warming trend since 2000; this demonstrates their inability to model even this fundamental and macroscopic natural oscillation of the climate system.

The existence of 20- and 60-year natural oscillations at the observed phases implies that about 50% of the warming observed from 1970 to 2000 is likely to have been caused naturally (Scafetta 2010; 2012b; 2013). These two oscillations can be understood in terms of the wobbling of the Sun relative to the centre of mass of the solar system (Scafetta 2010), which is shown in Figure 3.7. This record reveals major oscillations at about 20-year and 60-year intervals due to the combined orbits of Jupiter and Saturn. To understand these two oscillations one needs to realise that while it takes 11.86 years for Jupiter to orbit the Sun, and 29.45 years for Saturn, the time between two consecutive conjunctions of these planets is about 19.86 years. That is the time it takes these planets to come back into alignment relative to the Sun.

Moreover, every three conjunctions at about 60 years, occur when these two planets are located nearly in the same constellation. This is because the orbit of each planet is elliptical; thus a trigon of Jupiter–Saturn conjunctions generates a 60-year recurrence in the wobbling of the Sun with both the 20- and the 60-year oscillations consistently found among the climatic major oscillations (Scafetta 2010; 2013; 2014; 2016). The solar wobbling can thus be understood as a proxy of the major gravitational oscillation of the heliosphere and of the inter-planetary electromagnetic field, which may influence the incoming cosmic ray flux and interplanetary dust concentration (such as, Scafetta

Figure 3.7 Solar wobbling

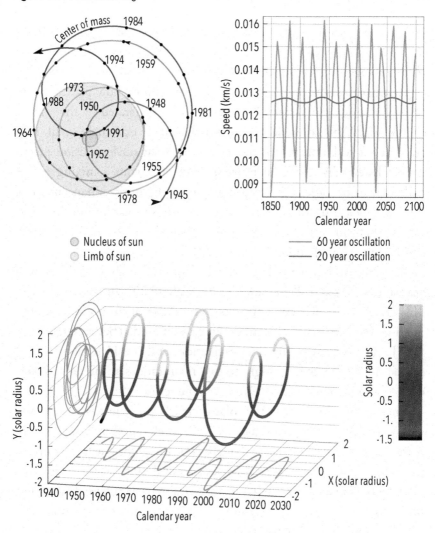

The wobbling (top left and bottom) and the speed (top right) of the Sun relative to the center of mass of the solar system. Highlighted are the main oscillations at 20 and 60 year periods due to Jupiter and Saturn's gravitational attraction.

2013, 2014; Kirby 2007; Svensmark et al. 2009, 2012; Ollila 2015). Scafetta et al. (2016) demonstrated that these harmonics are among the main stable resonances of the solar system.

Another possible physical mechanism explaining how planets could be modulating solar activity is the tidal forces generated by the planets on the Sun. While these forces are gravitationally extremely small, they are thought to be magnified by an internal solar nuclear fusion mechanism. Essentially, the tidal movement could favour a mixing of the hydrogen in the solar core, causing a small rhythmic enhancement of the solar nuclear fusion rate and, consequently, of solar luminosity. How this mechanism could be working is explained in more detail by Wolff and Patrone (2010) and by Scafetta (2012c).

The empirical evidence is provided by spectral analysis of the eleven-year sunspot cycle available since 1700 (Scafetta 2012d; Solheim 2013). This eleven-year solar cycle appears to be made of three oscillations: a quasi-eleven-year main cycle modulated and bounded between the 9.93-year Jupiter–Saturn spring tidal-oscillation, and the 11.86-year Jupiter orbital tidal-cycle, giving origin to two main modes at the ten- to eleven-year and the eleven- to twelve-year time scales.

Since 1850, on average, the solar cycle has been shorter than eleven years – the prevailing mode has been ten to eleven years long – which is the main reason why spectral analysis of the global surface temperature since 1850 gives a spectral peak at around 10.5 years, as shown in Figure 3.6. Planetary tidal frequencies have also been observed as changes in the total solar irradiance records – at short monthly and annual time scales (Hung 2007; Scafetta & Willson 2013b).

The finding that the sunspot cycle appears to be made of three main harmonics has been important in the development of a multiscale harmonic model for solar and climate cyclical variation throughout the Holocene, based on the two Jupiter–Saturn tidal frequencies, plus the eleven-year solar dynamo cycle (Scafetta 2012d; 2013).

Accepting this simple three-frequency model generates predictions of several other oscillations occurring in the climate system, such as the quasi-60-year, 115-year and 900- to 1000-year oscillations. Figure 3.8 shows the prediction of this model against a global climatic reconstruction for 2000 years (Ljungqvist 2010). It is, in fact, a good match at both the 100-year (secular) and 1000-year (millennial) scale.

Another possibility, which complements the above tidal mechanism, is that the movement of the planets around the Sun, by itself, alters and modulates the physical properties of the heliosphere. In fact, the planets not only generate a gravitational field but they are also surrounded by large magnetospheres that can interact with the solar wind, cosmic rays and the interplanetary dust-density distribution of the solar system, and electromagnetically link the entire solar system. It has been observed that

Figure 3.8 Modulated three-frequency harmonic model

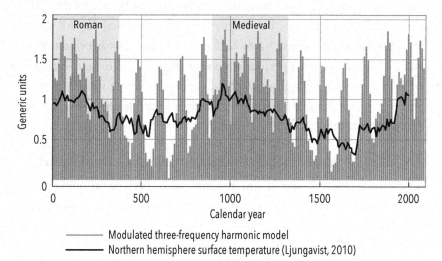

Source: Scafetta 2012d, and adapted with permission from Ljungqvist, FC, 'A new reconstruction of temperature variability in the extra-tropical Northern Hemisphere during the last two millennia', *Geografiska Annaler*, vol. 92, pp. 339-351, copyright 2010.

the side of the Sun facing Jupiter appears brighter (Scafetta & Willson 2013b); and numerous planetary harmonics are found in the aurora and meteorite historical records corresponding with this relationship (Scafetta 2012b; Scafetta & Willson 2013a).

Several other authors have hypothesised a possible planetary modulation of solar change causing climate change (for example, Abreu et al. 2012; McCracken et al. 2014; Morner 2013; Wilson 2013; Wolf 1859; and several others).

Finally, at shorter time scales, Wang et al. (2012) modelled several soli-lunar tidal waves as a function of the obliquity and revolution velocity of the Sun and the Moon, and of the latitude, radius, and rotation velocity of the Earth. These researchers demonstrated that such a rhythmic motion reproduces various recognisable climatic cycles including the 30- to 60-day oscillation, seasonality, and the El Niño–Southern Oscillation. I noted (Scafetta 2012a; 2014) that among these long-range, soli-lunar tidal oscillations there is a set linked to the 8.85-year period referring to the lunar apsidal line rotation period, and also to the nine-year period referring to half of the eighteen-year eclipse Saros cycle and of the 9.3-year period, which is half of the 18.6-year soli-lunar nodal cycle period (Haigh et al. 2011). These three periods average to about a 9.1-year period: a frequency which is found in the global surface temperature record, as shown in Figure 3.5.

Modelling climate change in terms of natural oscillations

The identification of these oscillations with periods at about 9.1, 10.5, 20, 60, 115, and 1000 years, and then their incorporation into a model, has enabled me to reconstruct global surface temperature change since 1850, as shown in Figure 3.9 (Scafetta 2013). In particular, the natural component of climate change based on the modelling of these oscillations is provided by the bottom green curve (six-frequency harmonic component alone). The grey curve represents the combined contribution of the volcano signal and of the anthropogenic signal. The latter is calculated using the corresponding radiative forcings, assuming that the climate sensitivity to them is about half (factors: $\beta=0.4$ to $\beta=0.6$) of that predicted by the CMIP5 GCMs used by the IPCC. The combination of the various natural and anthropogenic contributions gives the top green ($\beta=0.6$) and dark green ($\beta=0.4$) model curves. A comparison between these and the global surface temperature record highlights the good match at multiple time scales between the measured and modelled records.

Figure 3.9 Historic and projected temperature change

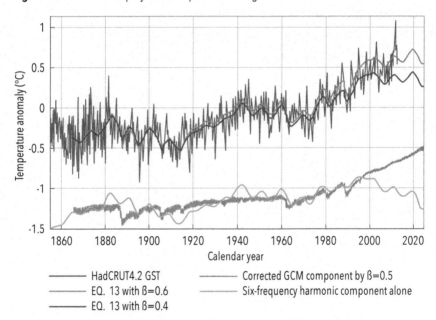

Modelling of homogenized global surface temperature record (red line) using the semi-empirical model (top green lines) based on natural oscillations (bottom light green), plus the volcano and anthropogenic contribution (grey line).

Source: Scafetta 2013.

The good match observed in Figure 3.9 is unparalleled by any climate model used by the IPCC (Scafetta 2013). This suggests that interpreting climate change through these oscillations, as demonstrated in this chapter, is the correct way to understand and combine temperature change due to natural and anthropogenic influences. It also provides a basis for making forecasts of future climate change. Because of natural oscillations, the real climate sensitivity to greenhouse gases is about half of that assumed by the current IPCC climate models. Corrected forecast models suggest that throughout the twenty-first century the climate could still be warming, but moderately – that is less than 2 °C (Scafetta 2013). All the evidence suggests that solar, lunar and planetary oscillations drive the natural oscillations observed in the climate.

4 The Role of the Moon in Weather Forecasting

Ken Ring

The Moon affects the weather around the planet. As the Moon orbits the earth it is responsible for deep ocean currents, which set the direction and force of the winds; and it forms and changes the intensities of anticyclones and depressions. Because of the Moon, we get changes in temperatures and barometric pressures, and in weather conditions – from cyclones to heat waves. The Moon also has a major role in the creation of volcanoes, the timing of their eruptions, and in earthquakes and tornadoes.

Back in the 1970s, when I lived for ten years on the wild east coast of the northern island of New Zealand, I began to suspect that the Moon influenced the weather. I ran fishing nets each day, working with the tides. Before long, I realised that with every king (or spring) tide, there was either an accompanying storm or, at least, over the higher-tide period, a noticeable change to unsettled weather. It seemed a reasonable deduction that it was not just happening in isolation there. The weather was either answerable or parallel to the vagaries of the ocean, and, likewise, the atmosphere seemed to have its own tide. By recording when the king tides were due, I was able to plan for inclement conditions well ahead, with some measure of success.

In hindsight, it was my good fortune to begin my investigations in 1975, because unbeknown to me the Moon was entering a period when its three main cycles were peaking and ebbing in combination, making the patterns more defined. It is the nature of cycles to ebb and flow with peaks and troughs.

In the course of its journeying around the Earth each month, the Moon has three main cycles: phase, perigee, and declination. Each cycle is of a different duration, and usually they are out of synch with each other; however, when two or more cycles peak or trough together in a month their combinations can be very influential. This is how the ocean's king tides occur.

Phase

The cycle most obvious to our sight is that of the full moon to new moon and back to the full moon, known as the phase cycle, which takes 29.5 days. What we see during this cycle is the changing angle of the Sun's illumination falling on the Moon's surface. The side of the Moon towards the Sun is lit, so *before* the full-moon phase its left side (for Southern Hemisphere viewers) is the lit side. The right side of the Moon is lit *after* the full-moon phase.

At the time of the new-moon phase the Moon is between the Earth and the Sun, while at the full moon it is the Earth that is in the middle. Gravitation increases when the Earth, Moon and Sun are all in alignment at the times of the new and full moons. Moon phase contributes to the timing of changes and intensity of weather rather than the type of weather conditions.

The Moon rises, on average, about 50 minutes later each day. It may be observed that a weather condition on one day may recur at about 50 minutes later the following day. For example, on the evening of a full moon, the Moon will rise at sunset and set at sunrise. Cloud cover typically recedes during that interval and resumes again after the Moon has set. But on successive days after the full moon the cloud cover may advance into the night hours, receding after midnight with the onset of the next phase, known as the last quarter.

The general rule is that rain, if about, is more likely when the Moon is gone from the sky. So when the Moon is visible, leave the umbrella at home. At new moon – a day moon – the Moon is overhead for much of the day, and rain is less likely. New moon days can be pleasantly dry. A new moon sets at dusk and is gone from the night sky, and if rain is about then it will typically fall during night hours. If no rain is about, during new moon after the Moon has set, temperatures may just drop. If it is raining during a new moon day, one should also expect rain that night.

During the first quarter phase, when the Moon rises at noon and sets at midnight, rain is mostly between midnight and noon. If it is dry before noon during the first quarter, it will most often also be dry during the afternoon.

During the last quarter phase, the Moon rises at midnight and sets at noon, and rain, if about, mostly comes between noon and midnight. But if rain falls before noon during the last quarter, even more rain will arrive during the afternoon and evening.

Perigee

While the phase cycle is easily seen, the 27.5-day perigee cycle is less so.

Astronomers know that in an elliptical orbit an object moves at varying speeds: slower at the point of the long axis, and faster at the point of the shorter axis. In the Moon's monthly orbit of the Earth, there is a day when the Moon is at its closest to Earth, known as the perigee of the Moon's orbit. Half a month later, the Moon is at its furthest point from earth, known as the apogee for that 27.5-day month.

In ancient times perigee was measured by means of a calibrated stick. Each day, holding the stick aloft and at arm's length against the Moon in the sky, the local priest or astronomer would slide his thumb along the stick to find the day when the diameter of the Moon was at its widest. This information was important for fishermen, and still holds today. Don't fish on perigee day, because the increased wave and surf turbulence deter the fish from coming in too close to shore. Two days before or after perigee is ideal. The lunar planting calendar also recommends not sowing crops at the time of perigee, because the soil is more electrically disturbed, affecting early germination.

Perigee creates the king tides. Because of the Moon's increased orbital speed relative to that of the Earth, perigee brings extra turbulence to land, sea and air – inducing earthquakes, maximum tides, and heavy storms – and with the potential for snow at low altitudes in winter, heat waves in summer, damaging floods, and the formation of cyclones.

For each month of every year, the perigee Earth-to-Moon distance varies, with closer perigees being more potent than others. In 2016, the closest perigee was on 14 November, at a distance of 356,509 km from Earth. The total perigee cycle is 8.85 years. From 2010 to 2015, perigee occurred over the Southern Hemisphere, bringing relatively more destructive weather there. For about the next four years, perigee occurs over the Northern Hemisphere. It changes back to the Southern Hemisphere in about 2020.

To modify the heat or cold of the seasons, the perigee, apogee, and the full or new moons work together. While it is the increasing or decreasing distance of the tilted Earth to the Sun that causes seasons to change annually, whether or not the summer or winter is warmer or cooler than average can be influenced by how these lunar factors combine. Summer full moons, when in perigee, can cause overall higher temperatures, and winter new moons in perigee can bring average lower minima. Generally, the stamp of perigee is the exaggeration of seasonal weather conditions, or of whatever weather pattern would have been unfolding anyway. Most extreme weather events will be found at or near perigee days, or midway between. When heavy rain accompanies perigee-related king tides, water cannot escape and flooding is often the result. In zones prone to earthquakes, there is twice the chance of significant tremors in the week straddling perigee, compared to the other weeks in a month, especially if this is at or near the full or new moon.

The Moon at perigee is said to increase tides by 20%. This can add another third to the average tidal height. Cyclone formation is greatest around summer perigee when the water surface is warmer, enabling more evaporation – particularly when it is accompanied by the full moon. Tropical cyclones (or hurricanes and typhoons), tropical lows and depressions, thunderstorms and strong gales can all be blamed on the Moon. For a cyclone to develop, the surface of the sea must reach a temperature of about 28 °C. This can only occur when the Sun is over that particular hemisphere and close to, or past, its solstice position for that tropic, which will allow sufficient time for the Sun's rays to heat the sea to that critical temperature. The heating process will continue after the Moon has crossed the Equator and into the other hemisphere.

The Moon moves relatively faster around the earth at three times: during new moon; when it is perigee; and when crossing the Equator. When all three coincide very rough conditions can result. This last occurred on 9–10 March and 7–8 April 2016. The apogee comes two weeks after perigee when the Moon is at its farthest for the month. The variation between the distance of the Moon from the Earth during perigee and the apogee is about 50,000 km. In 2016, the furthest apogee came on 31 October.

An apogee tends to bring reduced winds, and more beneficial conditions for newly germinated sprouts. Apogee is one of three days during an average month when the Moon slows; the other two are on the day of the full moon, and on the days of northern or southern declination. When all three of these coincide within three days, it brings stable weather.

Declination

There is a third cycle that lasts for a period of 27.3 days.

Ideally, the Moon would orbit the Earth around the Equator, but because the Earth is tilted between 21° and 23° to the perpendicular, the Moon must cross between one of the Earth's hemispheres to the other and back again within any month. This is called the declination cycle. It is this cycle that causes the Moon to move huge volumes of water across the Equator; it also changes barometric pressures, enabling a distribution of the atmosphere around the planet. Each month this crossing of the Equator circulates warm equatorial air outwards into the temperate zones, thereby increasing their temperatures.

Mostly, it is the declination cycle that determines the actual type of weather, whereas the other cycles adjust these incoming conditions for intensity and timing. When the Moon crosses the Equator, heading either north or south, it is rising due east and moving faster, akin to the midpoint of a pendulum. Also called the lunar equinox, it results in faster-moving weather systems, which we observe as more change-able weather. This happens twice a month: as the Moon crosses the Equator heading north; and thirteen to fourteen days later crossing the Equator heading south. The most common time for the barometer to drop is within two days of the lunar equinox in any month, and is the most common period for atmospheric disturbance.

When the Moon is at about the highest northern or the lowest southern declination point – observable as the moonrise at either far north or far south along the eastern horizon – weather will slow down and persist without too much change. Whatever the weather conditions, a day out from northern or southern declination it will usually remain unchanged until at least a day after it.

Airflows typically follow the Moon's direction. When the Moon rises at around the declination points, its orbit is parallel to that of the Earth, and weather maps will show winds being dragged in the direc-tion of latitude lines. But heading north after southern declination in the Southern Hemisphere, winds are generated towards the northeast, and as the Moon treks southward after northern declination it generates winds towards the southeast.

The situation reverses for the Northern Hemisphere, with winds towards southeast after northern declination and towards northeast after southern declination.

The Moon at the northern declination generates warm moist rain-bearing winds for the Southern Hemisphere and cold polar winds over the Northern Hemisphere. Southern declinations and the few days afterwards generate wintry blasts and winter snow dumps for New Zealand and Australian ski fields. The closer these high and low points are to the full moon during the winter of each hemisphere, the colder will be the wintry blasts. The distance a location is from the Equator modifies the conditions. Northern declination affects the heat of the north of Australia more than it does the southern half of Australia, just as the south of Australia and New Zealand are more cooled by the southern declination than the northern halves of both countries.

It is when the Moon is going through its lunar equinox that thunderstorms and electrical activity occur more frequently, while steady drawn-out rain occurs more when the Moon is at the north or south declination points. If a full or new moon and the north or south declination occur together, or about a day apart, more intense heat or cold will result. In June for the Northern Hemisphere, or December for the Southern Hemisphere, the tendency for weather systems to become stationary for a longer period is increased and even extended for a few days if the new or full moon falls two or three days either side of the solstice position; in other words, the two stationary periods may run into one another. In summer, with anticyclonic conditions, these circumstances may be the beginning of a prolonged dry period, or, in the winter of unsettled weather, and it could remain that way for a week or more. If an anticyclone is also residing at that time, then a prolonged spell of frosts may be likely.

Declination slowly changes over a period of 18.6 years. To this beat we get a grand recycling of seasonal weather trends. The average latitude of the Earth over which the Moon reaches is 23° north (N) to 23° south (S). This range increases by one degree a year until it reaches a maximum of 28° N and 28° S of the Equator, and this is called the 'major standstill'. The last major standstill was on 22 March 2006. Nine years later the minimum range or 'minor standstill' is reached, between 18° N and 18° S, the last of which was on 21 September 2015. In 2025, the Moon again reaches 28° N to 28° S, and this will be the next major standstill. The lack of understanding of such changes within lunar cycles can sometimes lead to a belief in changes in climate.

During major standstill years, when the Moon must cross hemispheres with greater angular speed, weather patterns change relatively more rapidly. During these years, the weather may be described as being more bizarre, or fickle, for example, during the years 1950, 1969, 1986 and 2006. Conversely, during the years of minor standstill, as the Moon crosses hemispheres at a more leisurely pace, the slower moving Moon brings weather patterns that take longer to develop to full strength, often lingering noticeably before declining, for example, in 1959, 1978, 1997 and 2015.

In normal circumstances an anticyclone lasts about two to three days. When the Moon's declination is at 18°, the anticyclones take longer to pass over the country, and troughs and depressions seem reluctant to leave. Longer frost periods can result. Even though weather systems are slow to move during new- or full-moon phases anyway, they will become even slower moving or even stationary around the minor standstill period.

This means that trends are somewhat predictable from the stages in the declination cycle. From about 2003, when the Moon was at 24° declination and up to 2006–7 when the Moon arrived at the 28° mark, the Moon was circulating the warm atmosphere to beyond the tropics and into the temperate areas. Hence, the temperature averages began rising from about 2002, just as they had done between 1984 and 1987.

Figure 4.1 Major and minor lunar standstill

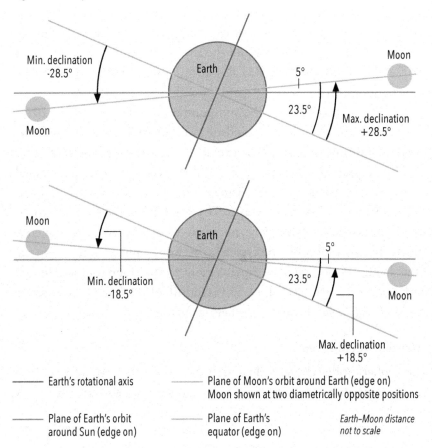

As the Earth rotates around the Sun, the Sun's declination changes from +23.5° at the northern hemisphere Summer Solstice to −23.5° at the northern hemisphere Winter Solstice. As a consequence, in the northern hemisphere the Sun is higher in the sky and visible for a longer period of time in June than it is in December. The Moon also changes in declination, but on a much shorter cycle of approximately 27 days. The range of this declination also changes because the plane of the Moon's orbit around the Earth is inclined by about 5° to the ecliptic (the plane of the Earth's orbit around the Sun). The direction of lunar orbit inclination gradually changes over an 18.6-year cycle, alternately adding to, or subtracting from, the 23.5° tilt of the Earth's axis. A major lunar standstill, as shown in the top chart, occurs when the moon changes its declination during a single nodal period from +28.5° to −28.5°. A minor lunar standstill will occur 9.3 years later with a reduced range during the single nodal period from +18.5° to −18.5°, as shown in the bottom chart.

Source: Matt, A 2011, 'Lunar Standstill.GIF', Wikimedia Commons, viewed 7 April 2017, https://commons.wikimedia.org/w/index.php?curid=40572823.

In about 2010, temperature averages began a downward slide, just as they had done at the same stage of the last declination cycle in 1990. In 2009–10 the sunspot cycle was also at minimum, which added to the cooler conditions.

In 2015 (and in 1996–7) the minimum declination was 18°. In 2016, one year on from the minor standstill, we moved away from the slower-moving weather patterns. In Australia, a longer rain-bearing depression is called a cut-off low; but in New Zealand the persistence of rain is blamed on large high-pressure systems sitting to the east, called a blocking high. In minor standstill years, such as in 2015 and 1997, the Moon is not as far south towards the South Pole, nor as far north for the Northern Hemisphere in winter months, and typically this brings milder winters. On the other hand, a major standstill causes a more northern and southern moon, and more severe winters. The major standstill of 2006 broke records for cold in New Zealand, and for dryness in Australia.

Atmospheric tides

Atmospheric tides ebb and flow each day over every part of the globe in the same manner as an ocean tide. The atmosphere extends higher into the heavens when the air tide is 'in', which is when the Moon is above the horizon. After the Moon has set, the air reduces in its height from ground level, and then the air tide is 'out'. The atmosphere's height can vary by up to 25% between moon phases. A tide in the air is not as visible as a sea tide, but can be logically deduced from observations. Sailors will tell of a big blow just before a turn of the sea tide, then, right on the turn, the wind will drop. If rain is about, that is when it invariably falls. Meteorologists observe that weather balloons float higher around new- or full-moon king tides, due to what must also be a king tide in the air. A ring of cloud encircling a high hill or mountain peak will ascend or descend in a tidal manner as the Moon itself rises in the sky, or sets. Cumulus cloud will only build up without a moon, and that cumulus cloud may be the forerunner to hail or lightning. Cloud tends to diminish more when the Moon is overhead.

The reason is that the Moon increases the volume of air beneath it. This air acts as an insulation layer, so the higher the atmosphere, the greater its insulating properties. A higher atmosphere in the daytime will prevent the extreme heat of the Sun from reaching ground level, and at night, higher air levels will keep the cold of space relatively further away from the ground. When the air height lowers, once the Moon has set and taken the air bulge with it, the cold of space can come closer to Earth and the subsequent drop in temperature can cause clouds to condense. This happens when the Moon is absent, which is during the day of a full moon, the night of a new moon, the morning of the first quarter, and the afternoon of the last quarter. It is why, at those times, if the possibility of rain is about, it will be more likely to fall. The buildup of heat to form a cyclone speeds up during a full moon and into the last quarter phase because that is the time for a low atmospheric tide, both from the Moon being a night moon and because the Moon is over the opposite hemisphere.

It is why around the time of full moon, which is only in the sky overnight, clouds mostly appear around noon. The atmosphere attains its greatest altitude on the night of a full moon. If there are no clouds and it is summer time, the heat of the Sun on the day of a full moon may be the most intense for that month, because there is less air to impede the Sun's heat from reaching the ground. Full-moon nights will nearly always be clear because the air is stretched higher and the resulting extra insulation prevents cooler heavier night air descending, rendering it less able to condense water vapour to create cloud.

The reason why we usually see the full moon in all its shining glory is because it makes us see it, by clearing the sky. In old weather lore there is a nautical saying: 'The full moon eats clouds'. If skies are clear during the day of the full moon, one can always expect a clear night, especially on either side of midnight. It would be illogical for the Moon to only control weather on the full-moon night. More likely it does it every night, but less obviously.

When the Moon is not visible in the sky – for example during the daytime of a full moon, and especially in the summer when the Moon is over the Northern Hemisphere – if a drifting cloud obscures the Sun the temperatures will suddenly cool. As the cloud drifts away the heat of the Sun will re-intensify. The explanation would be that the atmosphere is in a low-tide phase, and allows a greater amount of the Sun's heat to penetrate after the cloud has drifted away. This condition will also occur in the last quarter phase, during the afternoon, and in summer very high temperatures can occur. If a marathon is held at the time of a full moon, athletes are more likely to suffer heat stroke, as tragically happened in November 2011, in San Antonio.

Frosts in the winter may occur during any phase of the Moon, especially when there is an anticyclone over the country, but a frost will be heavier at new moon phase for two reasons. First, because the low-air tide will be at night, allowing the cold of space to fall closer to the ground. Cold air falls in the same manner that cold water finds its own level. Second, the new moon in winter is over the Northern Hemisphere, which assists in moving more of the atmosphere away from the Southern Hemisphere, creating a thinner density of air and increasing the effect of a low tide. If the Moon is in perigee, the atmospheric tide will be even lower than usual; the air tide will go 'out' more and, for example, during new-moon winter days, that lowered air tide will allow for even colder conditions, and hotter perigee full-moon summer days.

Propeller-driven light top-dressing aircraft may be more at risk taking off under load when the air is 'thin'. The day may start while the atmospheric tide is high with a morning moon at last quarter phase, which may have been a comparatively safe time to exceed normal load limits.

But after a lunch break, and after the Moon has set, the atmospheric tide will be low and thin. The plane may require extra propulsion to rise quickly enough to clear possible obstructions, such as trees, buildings or small hills a short distance away. For the same reason, on a full-moon day, racing yachts often need the assistance of a greater surface area of sail, for example, a spinnaker. There will be thinner air during the night of the new moon, the morning of the first quarter, around midday on the day of a full moon, and during the afternoon of the last quarter.

I spoke with the late Sir Edmund Hilary, conqueror of Mount Everest; he was interested in why it was sometimes possible to breathe at 16,000 feet and not at other times, and had a personal hunch that it was weather and, possibly, moon-related.

In northern China people are told not to climb high mountains before a new moon. Tibetans prefer to climb between the new moon and the first quarter. All recorded oxygen-less ascents have been in this time window.

When climbing the world's highest mountains, more fatalities occur on descents than on ascents. Climbers are more fatigued and take less care. Descents are more physically taxing than ascents and any oxygen depletion can be dangerous. If descending in late afternoon or early evening, a climber would want plenty of naturally available air. But the relatively greater volumes of oxygen sitting at higher levels in the afternoon are only available for two or three days after the new moon.

As mentioned, the Moon at perigee can add a third to the normal height of a sea tide. It can also be shown that over time trends in air pressures will rise and fall in proportion to ocean height. The aggregated differences between very low and very high sea tides at perigee have not been calculated, but whatever the difference is, if this parallel was made for atmospheric tides, the difference between an atmospheric low and an atmospheric high tide would be significant.

A long-range forecasting tool

At corresponding days that are nineteen years apart, the Moon has approximately the same phase and occupies approximately the same position in the sky with respect to the background stars. The tropical year is approximately 365.242 days, and one synodic month (lunation) is approximately 29.53 days. But:

> 19 years = 6939.60 days and
> 235 lunations = 6939.69 days.

So, a period of nineteen years (known to the Greeks as the Metonic cycle) differs from 235 lunations by only 0.09 days, that is, around two hours and ten minutes. Today's calendar is solar-based, and yet, remarkably, the Moon's cycle and that of the Sun come into synch for one day every 19 of these solar-based years. To the day, seven Metonic cycles bring the Milky Way, as observed from Earth, exactly back into line. It does not end there, because also, to the day, ten Metonic cycles bring a return of the relative positions of the inner planets of the cosmos.

Long-range forecasting is possible when similar weather conditions recycle. If it is accepted that the Moon can change or control weather, then when Moon positions repeat so should observable weather patterns. Therefore, the nineteen-year intervals should work more often than not. In ancient stone circles, such as at Stonehenge, we find circles with nineteen markers. There are many such structures in countries as far apart as Australia and Ireland, which suggest that these monuments may have been long-range weather calculators.

But if visual proof is needed, it will be found that the general appearance of a weather map from exactly nineteen years ago will approximately repeat itself. So, this cycle is one of several that we employ for the weather almanacs we produce for Australia, New Zealand, and Ireland.

Figure 4.2 Comparing weather maps

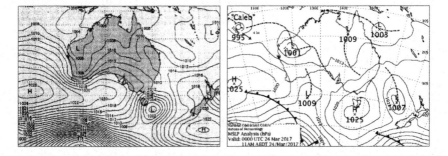

The weather map on the left was published in Ken Ring's Australian Weather Almanac as his prediction for 24 March 2017. The map on the right shows actual mean sea level pressures for this date, as record and subsequently archived by the Australian Bureau of Meteorology. The low off the north-east coast of Australia, shown in both weather maps, developed into Tropical Cyclone Debbie.

Source: *Australia Weather Almanac 2017*, Ken Ring Ltd, Auckland, copyright 2016; and Australian Bureau of Meteorology website (http://www.bom.gov.au/australia/charts/archive/) reprinted under Creative Commons (CC) Attribution 3.0 licence.

Observers sometimes remark that bands of anticyclones or depressions seem to be passing across further north or south than normal for the time of year. The explanation is that such bands may be at corresponding latitudes over a nineteen-year return. In the winter months, this situation may become more noticeable when an anticyclone is slow moving through the Australian Bight and over the Tasman Sea, and a southerly airflow affects the lower half of Australia and New Zealand, giving cold bleak weather with snow to low altitudes over some areas, just as it did one or more nineteen-year cycles ago.

The Metonic and other lunar cycles have so far been ignored by mainstream meteorology. It is an area waiting to be properly researched. Meteorological models use averages, the disadvantage being that averages are based on solar months. This may work for seasonal temperatures, which come primarily from the input of the Sun, but for the timing of winds and precipitation, cyclones, tornadoes and other extreme events, better results may be obtained if the months being averaged are converted to lunar ones. This process will permit the phases of the Moon to be factored in.

Weather systems advance in a generally east direction at various speeds. There may be little or no movement for many days at a time, or rapid or slow changes. Over an eighteen- to twenty-year period, what changes weather the most is the declination cycle. Because of lunar declination wind directions change together with barometric pressures. Perigee can bring changes in intensity, creating extreme events such as gales and storms when normally there would just be moderate levels of wind. The phase cycle of the full to new moon provides changes in daily temperatures and the timing of rain.

As we understand weather, it is these variations that bring changing conditions, and the cumulative effects of weather become the climate.

5 Creating a False Warming Signal in the US Temperature Record

Anthony Watts

In climate science, the metric by which change is gauged is the tracking of a global surface temperature value over time, usually in periods of 30 years or more, which is a period of time that is said to define climate for a given area of the Earth. However, there is no universally accepted definition for Earth's average temperature, and, in fact, several different groups around the world use different methods for tracking the global average over time.

It can also be said that all the key institutions apply adjustments to the actual measurements, which has the effect of exaggerating, rather than correcting for, the urban heat island (UHI) effect.[1] I do not dispute that many of these adjustments are justified. However, in this chapter I show that the overall process, as currently implemented in the United States, causes more problems that it fixes.

If this were forensic science, such polluted and corrupted data would be tossed out as being unsuitable for the purpose of making a legal decision. Yet entire economies and national policies are being modified

---

1   UHI, also known as the 'urban heat island', is the tendency for cities to be warmer because they retain the daytime solar radiation in the concrete, brick, and asphalt infrastructure, and then release that heat at night as infrared, thus warming the air that would normally cool.

based on the trends seen in this data, as well as the future projections. When science refuses to own up to bad data, it is no longer science.

How temperature data is recorded and transmitted

To obtain a global surface temperature value, thousands of daily high and low temperature readings are taken around the world and submitted to a database. But, these numbers often contain errors, and a number of adjustments are made to this original raw data to make an accurate estimate of global average temperature possible and reasonable.

Consider observers of daily weather records, who volunteer as part of the US Cooperative Observer Program (COOP) with the National Oceanic and Atmospheric Administration (NOAA). These can often be people who, for example, run weather stations provided by the NOAA on their farms or even in their backyards. In other cases, these weather stations could be fire stations, police stations, ranger stations, water pumping plants, or even sewage treatment plants. In the US, the COOP makes use of data from all these sources.

The reason all these sources are used is because, as with any human endeavour, there can be problems. In the past, the daily temperature record was done visually, using thermometers housed in shelters, and the results recorded as the daily high and low temperature on a piece of paper – the B91 form (National Weather Service 2006). Once a month, these B91 forms were mailed to the National Climatic Data Center (NCDC) in Asheville, North Carolina, where they were transcribed and collated into a numerical record representing the high and low temperatures of each day of the month. From these the monthly global average temperature is calculated.

However, people get sick, go on vacation, or are just simply negligent because they are preoccupied with other commitments. Consequently, the temperature record can have gaps or errors. An inspection of B91 forms shows missing days of data (National Climatic Data Center 2007), improperly recorded values, and transcription errors (Shein 2008).

Imagine those issues on a national or even global scale and you can appreciate the need to make adjustments and corrections to fix the data and to make it usable when determining the best estimate of global average temperature.

The procedures used to process and adjust climate station data are shown in Figure 5.1. This methodology is applied to data whether it is from urban sites, sites that have been moved, or from what I will refer to as 'pristine' locations where temperatures have been recorded by a single person, at the one site, using the same equipment, from the beginning of the record (Goetz 2008).

Figure 5.1 Procedures used to process and adjust temperatures

| Read min/max temperatures | Record temps and observation time in B91 form | Once a month mail to NOAA | NOAA converts F to C and logs result in station's .dly file | Check quality and eliminate gross errors - but first convert back to F! | Calculate monthly averages from min/max |
| A monthly average can often be found when some min/max are missing! | Store the Areal data in USHCN and GHCN | Adjust for time of observation (TOB) | Apply the USHCN Version 2 homogenisation algorithm | Eliminate missing values using FILNET | Store the TOB and FILNET results with the Areal data in USHCN |

Source: Goetz 2008.

In a *New York Times* article about an ostensibly 'pristine' site in New York state, temperatures were purported to have increased by 2.7 °F in 112 years (DePalma 2008):

> Mr. Huth opened the weather station, a louvred box about the size of a suitcase, and leaned in. He checked the high and low temperatures of the day on a pair of official Weather Service thermometers and then manually reset them …
>
> If the procedure seems old-fashioned, that is just as it is intended. The temperatures that Mr. Huth recorded that day were the 41,152nd daily readings at this station, each taken exactly the same way. 'Sometimes it feels like I've done most of them myself,' said Mr. Huth, who is one of only five people to have served as official weather observer at this station since the first reading was taken on Jan. 1, 1896.
>
> That extremely limited number of observers greatly enhances the reliability, and therefore the value, of the data. Other weather stations have operated longer, but few match Mohonk's consistency and reliability. 'The quality of their observations is second to none on a number of counts,' said Raymond G. O'Keefe, a meteorologist at the National Weather Service office in Albany. 'They're very precise, they keep great records and they've done it for a very long time.'
>
> Mohonk's data stands apart from that of most other cooperative weather observers in other respects as well. The station has never been moved, and the resort, along with the area immediately surrounding the box, has hardly changed over time.

The data collected at this site could be considered relatively high quality. It was recorded by just five observers who were committed to their work. According to Mr Huth, there have been no station moves, and no equipment changes.

Surprisingly, this 'pristine' temperature recording station is only 'pristine' because of its long record of uninterrupted daily readings. The site itself has been affected in many ways. On-site inspection shows that the thermometer shelter is lower to the ground than is standard, resulting in a warmer temperature record. In addition, it is near the chimney of a nearby building and within the shade zone of a nearby tree, which makes daytime highs cooler, and night-time lows warmer in the summer, as shown in Figure 5.2. It is also wind-shaded by the building and by bushes, which also bias the night-time temperature upwards by preventing a full mixing of the air near the station.

Despite the Mohonk temperature records being of purported high quality, NOAA and the Goddard Institute for Space Studies (GISS),

Figure 5.2 Location of the Mohonk weather station, New York

Location of the Mohonk weather station (white box) housing weather recording equipment. The top photograph shows its proximity to a chimney, the bottom photograph is from a different angle.

Source: Map data from Google, DigitalGlobe. Photograph taken by Evan Jones.

which is part of the National Aeronautics and Space Administration (NASA), make adjustments to the data measured there. These adjustments add to the uncertainty of the real temperature record because they are applied by an automated process, without knowledge of the actual conditions at the Mohonk weather station.

How temperature data is adjusted

First, the temperature data is adjusted for the time of observation (TOB) and stored as a separate entry in a file called *hcn_doe_mean_data*. TOB adjustment is briefly described by NOAA's National Climatic Data Center (2012a). However, the actual methodology is such that it cannot be fully replicated outside these organisations.

Following the TOB adjustment, the series is tested for homogeneity. This procedure evaluates non-climatic discontinuities (artificial change points) in a station's temperature caused by random changes to a station, such as equipment relocations and changes. The version 2 homogeneity algorithm (National Oceanic and Atmospheric Administration 2012b) considers up to 40 highly-correlated series from nearby stations. The result of this homogenisation is then passed on to another algorithm called FILNET (National Oceanic and Atmospheric Administration 2012c), which creates estimates for any missing data by using data from nearby stations.

The NOAA's 'homogeneity algorithm' is an important step in the process, because it is used to 'smooth' data for a single station by using 40 nearby stations. Likewise, each of the other stations nearby has its data adjusted, in part, by the results of the single station in question, which in this case is the 'pristine' Mohonk station.

The result of all this is that station data becomes well mixed with other station data in a way that biases it, depending on how the data from the other 40 stations appears. If the other 40 stations are warmer than the Mohonk station for the matching time frames, the Mohonk station will be adjusted upwards. If the other 40 stations average cooler, the Mohonk station data will be cooled for the matching time frames.

Essentially, you end up with a well-blended 'temperature smoothie', where, by the time it ends up in the global temperature database, no station retains its original data character.

That final 'digital file cabinet', as shown in Figure 5.1, is where the daily data ends up at the the US Historical Climatology Network (USHCN) database, after being processed by this entire stream of actions. From here, the data is used with thousands of other station records to calculate the US national average temperature, and the global average temperature, which NOAA issues monthly in its *State of the Climate Report*.

But the adjustments don't stop there. Other agencies, such as the Goddard Institute for Space Studies (GISS) – formerly led by Dr James Hansen, described by some as the 'father of global warming' – add their own 'special sauce' to the mix of adjustments already done by NOAA.

Examples of urban heat island

One of the most common criticisms of the surface temperature record is that it has been slowly polluted over the last 100 years by the increasing urbanisation of our planet. A prime example is the Chicago O'Hare International Airport. In 2014 O'Hare became the busiest airport in the world.[2]

Chicago O'Hare has the International Civil Aviation Organization identifier of 'ORD'. Many people think that ORD stands for O'Hare, but that's not the case at all. The original name of the airport was not O'Hare, but rather Orchard Place/Douglas Field, and subsequently,

---

2 The busiest airports are measured by total movements with data provided by Airports Council International.

Orchard Field Airport. This is the source of its three-letter code, ORD. The airport wasn't renamed O'Hare until 1949.

Early photographs (ORD – A casual history of O'Hare Field, Chicago n.d.) show that Orchard Field was surrounded by farmland for several miles. Essentially, it was a rural airport, which expanded to become the megaplex of concrete and asphalt runways and tarmacs that we are familiar with today – the busiest airport in the world in 2014.

Just as the weather conditions (including surface temperature) were measured for aviation purposes back in 1942, so it is today. Yet the entire character of the land surrounding the weather station where the thermometer is located has dramatically changed. Gone are the rural fields surrounding the airport, replaced instead by airport and city infrastructure. Not surprisingly, data provided by NASA's GISS shows that there has been a dramatic increase in annual average temperature from the first complete year of temperature data in 1961 to the last complete year in 2013, as shown in Figure 5.3.

While there is no surprise that temperatures have gone up during the 52 years of available data, there remain these questions:

1. What part of the warming is attributable to greenhouse-gas increase, specifically carbon dioxide (CO_2)?
2. What part of the warming is due to natural variation?
3. What part of the warming is due to waste heat from human energy use activities?
4. What part of the warming is due to effects observed from land-use change, such as having more impermeable heat-retaining surfaces near the thermometer?
5. Since both NOAA and NASA highly adjust the data to provide end-products for public consumption and policy making, what portion of the warming trend is rooted in statistical methodology?

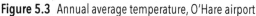

Figure 5.3 Annual average temperature, O'Hare airport

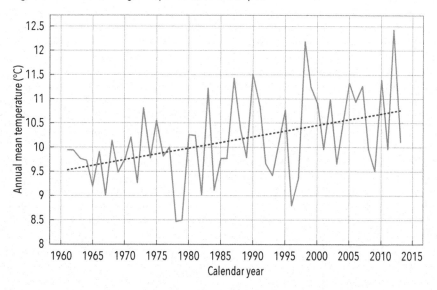

Measured temperatures have increased 2.35 °C per century for the period 1961-2013 as O'Hare airport has developed from a rural airport, to one of the busiest in the world.

Source: https://data.giss.nasa.gov/gistemp/.

When reading regular press releases and pronouncements from NOAA and GISS, one almost never sees anything other than 'climate change' being attributed to the observed increased level of CO_2. Any other factors are, essentially, ignored, or claims are made that they are insignificant factors, or that they have been dealt with by using statistical adjustments.

If we consider another example, the city of Las Vegas, Nevada, we see a similar temperature trend to that at Chicago O'Hare Airport, as shown in Figure 5.4.

GISS only shows one part of the temperature record – the average annual temperature. But, in fact, it is actually made up of two parts: the daily maximum temperature, and the daily minimum temperature.

Figure 5.4 Annual average temperature Las Vegas, Nevada

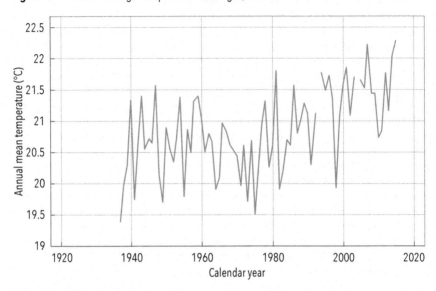

Source: https://data.giss.nasa.gov/gistemp/.

Figure 5.5 shows the temperatures as provided by the National Weather Service Office in Las Vegas, in *The Climate of Las Vegas* (Stachelski & Gorelow 2016). Here all three, the average, the minimum, and the maximum, are shown.

There has been an overall warming trend in average temperature for Las Vegas, as shown in Figure 5.5 (top chart). But, it turns out that most of that trend is in overnight minimum temperatures, as shown in the middle chart (see Figure 5.5). Minimum temperatures are most affected by the explosive growth in the population and infrastructure of Las Vegas and the resultant UHI. This has been noted by city planners (City of Las Vegas, Office of Sustainability 2010):

> The average temperature (measured at McCarran Airport) has risen four degrees in just four decades (1970's – 2000's) (sic). The largest temperature increase corresponds to the largest increase in population, with over one

Figure 5.5 Annual average, minimum and maximum temperatures Las Vegas, Nevada

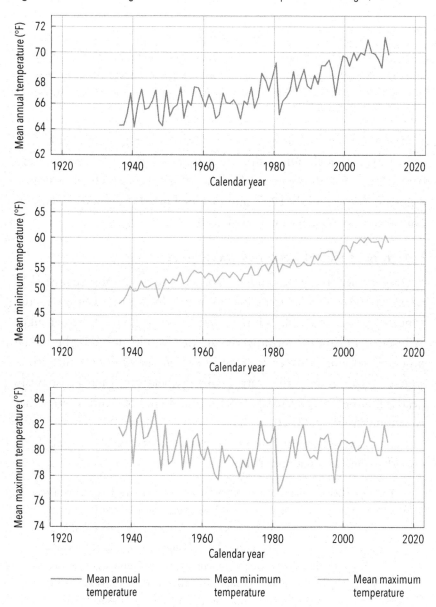

The average temperature shown in degree Fahrenheit (top chart) is calculated from the maximum (bottom chart) and minimum (middle chart) temperatures recorded at Las Vegas since 1937.

Source: Stachelski & Gorelow 2016.

million people relocating to the Las Vegas Valley during the same time-frame. As the population increased so did the demand for new infrastructure (roads & highways) and structures (residential & commercial). These new impervious surfaces absorb and radiate heat, resulting in an increase in the average temperature.

There has been no annual increase in record high temperatures, as shown by the maximum temperature trend in Figure 5.5 (see bottom chart). Since records began in 1937, the 117 °F record high temperature, originally set on 24 July 1942, and matched in 2005 and 2013, remains unbroken (National Weather Service 2016).

Urban heat island becomes waste heat

A UHI is where there is a tendency for cities to be warmer because they retain the heat of solar radiation absorbed during the daytime in the concrete, brick, and asphalt infrastructure; this is then released at night as infrared radiation, warming the air that would normally be in the process of cooling. This tends to increase the night-time air temperature, which, in turn, elevates the average daily temperature. There is also waste heat from the energy used for heating and cooling within cities that contributes to the UHI effect.

A study on waste heat (Murray & Heggie 2016) has shown that UHI can affect climate on a national level:

It has long been known that within large cities, thermal emission from heated buildings, industry and transport can contribute to a microclimate up to 12 °C warmer than background levels in the surrounding area, a phenomenon known as the Urban Heat Island (UHI) effect. Our results are strong evidence that changes in energy consumption contribute to temperature change over sub-decadal timescales in the two nations considered. Britain has experienced a drop in temperature of about 0.5 °C since the early years of the millennium at a time when world temperatures have remained virtually stable, whereas Japan experienced a rise in Δt of 1.0 °C between the early 1980s and 2000 (Fig.2, upper left), double the world rise in temperature over the same period.

Both these changes reflect changes in energy consumption in each country.

UHI effects can range from less than 1 °C up to the 12 °C, as cited by Murray and Heggie (2016). It doesn't matter if the city is in a warm, moderate, or cold climate, UHI is part of the nature of a city. Even at the small city level, waste heat can make itself known in the climate signal. For example, tiny Barrow, in Alaska, population 4373 as of 2013, is a town that could not exist without imported energy to keep the population from freezing to death during the long harsh winters.

The waste heat from Barrow has been clearly identified in the temperature record as a strong UHI effect. Hinkel et al. (2003), in 'The Urban Heat Island in Winter at Barrow, Alaska' say, 'The urban area averaged 2.2 °C warmer than the hinterland'. Yet, despite this obvious UHI effect on the temperature record, *Smithsonian Magazine* labelled it 'Ground Zero for Climate Change' (Reiss 2010).

Now, imagine this effect – of infrastructure encroachment and waste heat loss into the environment – being repeated worldwide over the last century, and, of course, being recorded as part of the climate data. World population has increased from about 2.5 billion in 1950, to more than 7 billion in 2012 (United States Census Bureau 2016). More infrastructure is needed to support this increase in population, yet the places where we measure our climate remain in the same places as population and infrastructure builds around them. As more of these heat-trapping surfaces are built, the more they absorb and radiate heat, especially at night, thereby affecting the minimum temperature, which, in turn, affects the average temperature.

My research (Watts 2015) has demonstrated that, at least for the US, this effect is widespread and can be separated from the climate record only by locating the few remaining long-term weather stations that remain undisturbed by encroaching infrastructure or other changes in the measurement environment at the station. Of the thousands of COOP stations that NOAA uses, only a subset of these are unperturbed stations that can give an accurate climate signal for the past 30 years (ibid.):

> Using NOAA's U.S. Historical Climatology Network, which comprises 1218 weather stations in the CONUS [Continental US], the researchers were able to identify a 410 station subset of 'unperturbed' stations that have not been moved, had equipment changes, or changes in time of observations, and thus require no 'adjustments' to their temperature record to account for these problems. The study focuses on finding trend differences between well sited and poorly sited weather stations, based on a WMO approved metric Leroy (2010) for classification and assessment of the quality of the measurements based on proximity to artificial heat sources and heat sinks which affect temperature measurement.

According to my study, bias at the micro-site level (the immediate environment of the sensor) in the unperturbed subset of USHCN stations has a significant effect on the mean (or average) temperature trend. Well-sited stations show significantly less warming from 1979 to 2008. These differences are significant with respect to the mean temperature, and most pronounced in the minimum temperature data.

Adjustments exaggerate urban heat island and waste heat contamination

The majority of weather stations used by NOAA to detect climate change temperature signals have been compromised by UHI and, or, micro-site biases. This affects temperature trends, and neither GISS nor NOAA's methods are correcting for this problem. The result is an inflated temperature trend that needs to be corrected.

In fact, NOAA and NASA's adjustment and homogenisation algorithms, which are applied to the data, actually increase the temperature trend. The 30-year mean temperature trend of unperturbed, well-sited stations is significantly lower than the mean temperature trend of the NOAA's NCDC official, adjusted, homogenised surface temperature record for all 1218 United States Historical Climatology Network (USHCN) stations (Watts 2015), as shown in Figure 5.6.

Based on surveys of weather stations across the continental US, less than 10% of the USHCN are free of biases associated with UHI, microsite bias, and other factors. Therefore, most US weather stations used to monitor climate are, in fact, unsuitable for the task.

Figure 5.6 Temperature trends for compliant, non-compliant and adjusted weather stations

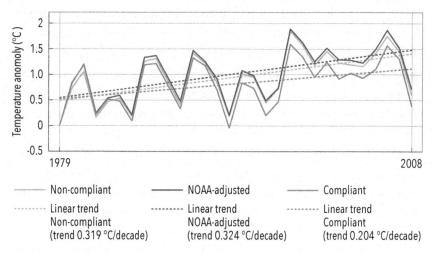

Weather stations across the continental US can be grouped based on their compliance with World Meteorological Organisation criteria. For the 30-year period 1979-2008, compliant stations (green) show a temperature increase less than the increase at adjusted (red) and non-compliant stations (orange).

Source: Watts 2015.

The NOAA's NCDC homogenisation adjustments cause well-sited stations to be adjusted upwards to match the trends of poorly sited stations. This increases the warming trend.

There is anecdotal evidence (Watts 2009) of the same siting problems occurring around the world at many other official weather stations, suggesting that the same upward bias on trend also manifests itself in the global temperature record. Homogenisation of the data, as currently performed in the US, actually makes the problem worse, and inflates the trend further.

Infilling, a key issue

Another key issue is the infilling of missing temperature data by drawing from the larger set of stations nearby – referred to in Figure 5.1 as estimating missing values. If only a few unperturbed stations remain in the entire surface temperature record, then their signal is small compared to the much greater population of perturbed stations that suffer from the effects of infrastructure encroachment. However, the data is all in-filled and blended together, and so the real climate signal from the unperturbed stations gets lost in that much greater set of noisy and warmer data. Effectively, the real long-term trend gets swamped by the greater population of warmer data.

This is not unlike how a drop of cold water in a full glass of water would be perceived. If the unperturbed stations were the drop of cold water, a person could easily detect the temperature of that water if the single drop were placed on the tongue. But if that same drop of cold water were placed in a glass of warmer water, the cold water becomes diluted and well mixed, and the small amount of cold water would be undetectable by the palate, even if the entire glass was consumed.

Conclusion

The homogenisation process can be theoretically justified given the problems that exist in many datasets. However, its method of application – summarised in Figure 5.1 – creates an inflated long-term temperature trend that is significantly warmer than data from the subset of unperturbed weather stations. This is evident for weather stations across the continent US (Watts 2015).

A solution would be to calculate temperature trends from unperturbed stations – specifically stations that have a long record, no moves, no equipment changes, no time-of-observation changes, and remain free of nearby infrastructure encroachment. Choosing only stations like this ensures that there is no need for adjustment of data, and that this data is representative of the true changes in the surface temperature over time.

Until the existing data quality problem is fixed, which has created an artificial warming bias, it is nonsense for the mass media to promote the idea of any year being the 'warmest year on record' (Miller 2016).

6 It was Hot in the USA – in the 1930s

Tony Heller & Dr Jennifer Marohasy

The United States' National Aeronautics and Space Administration (NASA) is an independent agency of the executive branch of the US federal government. It is responsible for both the Johnson Space Center in Houston, and the Goddard Institute for Space Studies (GISS) in New York. GISS publishes charts showing temperature change for both the continental US, and also for the entire Earth. Through its former director James Hansen, and current director Gavin Schmidt, the institution has also been an advocate for anthropogenic global warming theory.

In 2012, this advocacy caused a group of 49 former NASA employees – including seven astronauts formerly associated with the Johnson Space Center – to protest. They wrote an open letter to NASA headquarters in Washington (*Business Insider* 2012). It explained that the unbridled advocacy of carbon dioxide (CO_2) being the major cause of climate change is unbecoming of NASA's history of making objective assessments of all available scientific data prior to making decisions or public statements.

In this chapter, we show the extent of NASA's advocacy by contrasting actual measurements of hot days, with claims made by GISS NASA. In reality, the frequency, extent, and duration of US heat waves have declined sharply since the 1930s.

Source of data, and a key reason for adjustments

The US Historical Climatology Network (USHCN) is one of the most geographically and temporally complete temperature networks of any large area on the Earth. It consists of a network of about 1200 thermometers, which record the minimum and maximum temperature every day. The stations are relatively evenly spaced around the contiguous 48 US states. USHCN is an invaluable asset for understanding climate cycles in the US and for much of the Northern Hemisphere, because of the completeness and continuity of the dataset, as well as the proximity of the US to both the Atlantic and Pacific Oceans. The data from this network is compiled by the National Climatic Data Center (NCDC) in Asheville, North Carolina, which is part of the National Oceanic and Atmospheric Administration (NOAA).

NOAA makes adjustments to the USHCN temperature data, and then passes the adjusted data along to GISS, which then makes additional adjustments. GISS compiles the US data, and also data from other parts of the world, into a global dataset. This dataset informs the United Nations' (UN) Intergovernmental Panel on Climate Change (IPCC) deliberations.

According to GISS, one of the reasons why it is necessary to change historic temperature data relates to the time of observation (TOB). These TOBs adjustments are based on the theory that many station observers used to reset their minimum and maximum thermometers once per day in the afternoon, causing double counting of hot days. For example, on day one the temperature is 35 °C when the minimum and maximum thermometer is reset. It will still measure 35 °C that afternoon immediately after resetting. A cold front comes through that evening, and the second day is much cooler. But when the observer comes back to the weather station on the second day, the maximum temperature for the day is recorded as 35 °C, even though the actual maximum that day was 18 °C. Potentially, this is a serious problem, and if it were actually occurring would significantly affect the temperature record and cause past recorded temperatures to be too warm. It would be a problem of double counting on hot days.

One way to test the impact of TOB is to compare nearby stations with different observation times. According to NOAA records, from the 1930s through to the 1950s, Farmington, Missouri reset their thermometer at 7 am, while Warrenton, Missouri reset their thermometer at 5 pm. If TOB theory was correct, summer temperatures at Warrenton would be much hotter than at Farmington. This is not the case, as shown by the time series from Farmington and Warrenton, Missouri, in Figure 6.1.

Figure 6.1 Summer maximum temperature at Farmington and Warrenton, Missouri

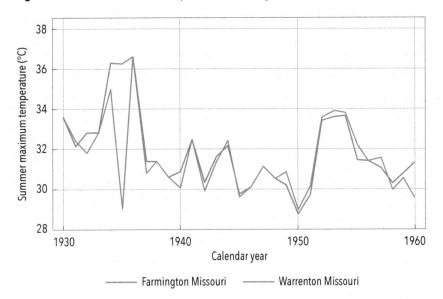

Source: NOAA GHCN Daily Temperature Record, June 2016, ftp://ftp.ncdc.noaa.gov/pub/data/ghcn/daily/all.

Heat waves in the USA

Considering the unadjusted data from US stations, the most notable change that has occurred in the US climate over the last 80 years is the sharp drop in the frequency, extent, and intensity of summer heat waves.

Since the 1930s, the average percent of days reaching 35 °C has nearly halved, as shown in Figure 6.2. This statistic is calculated by counting the number of days with maximum temperature readings over 35 °C at all of the stations each year, dividing the count by the number of maximum temperatures recorded at all stations during that year, and then multiplying by 100.

During the 1930s, about 7% of the daily maximum temperature readings were over 35 °C, but since 1960 that number is less than 5%.

Figure 6.2 Average percent of days over 35 °C

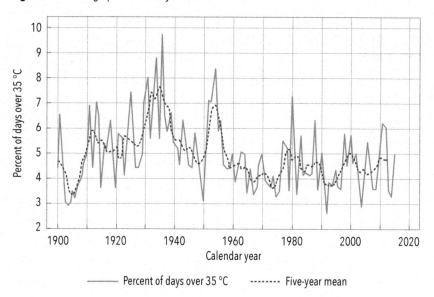

Source: NOAA GHCN Daily Temperature Record, June 2016, ftp://ftp.ncdc.noaa.gov/pub/data/ghcn/daily/all.

Figure 6.3 Annual percent of stations to reach 35 °C

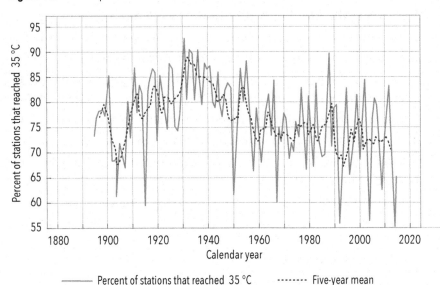

Source: NOAA GHCN Daily Temperature Record, June 2016, ftp://ftp.ncdc.noaa.gov/pub/data/ghcn/daily/all.

The area affected by hot weather in the US has also decreased. During the 1930s almost 90% of stations would reach 35 °C sometime during the year, as shown in Figure 6.3. Now just more than 70% of stations reach 35 °C each year. In 1936, nearly 25% of US stations reached 43 °C, but now that number is closer to 1%.

The duration of hot days has also declined sharply in the US. The US Environmental Protection Authority (EPA) publishes an Annual Heat Wave Index for the period from 1895 through to 2015, as shown in Figure 6.4. Consistent with our analysis of the USHCN data, the EPA chart also shows a spike in heatwaves during the 1930s. Clearly, it was hot in the US back in the 1930s – much hotter than during recent decades.

Figure 6.4 Annual Heat Wave Index, 1895-2015

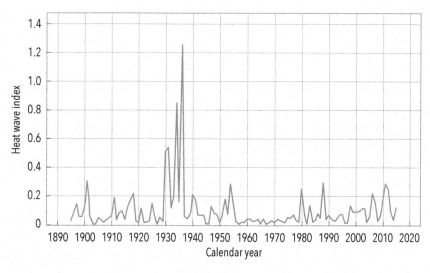

Source: EPA 2016.

The data for the US Heat Wave Index from 1895 to 2015 in Figure 6.4, shows data covering the contiguous 48 states. An index value of 0.2 (for example) could mean that 20% of the country experienced one heat wave, 10% of the country experienced two heat waves, or some other combination of frequency and area resulted in this value.

Changing historical temperature profiles

Up until the last few years GISS charts showed the 1930s as being relatively warm, and subsequently cooling up until 1970. For example, in 1999 Dr Hansen was the leading author of an article with a chart that showed nearly 0.5 °C cooling in the US from the mid-1930s through to the late 1990s – with 1934 more than 0.5 °C warmer than in 1998 (Hansen et al. 1999). In this temperature reconstruction, as shown in Figure 6.5, the year 1998 was only the fifth warmest year, with the 1930s being the warmest decade. The article included the following comment:

Empirical evidence does not lend much support to the notion that climate is headed precipitately towards more extreme heat and drought …

… in the U.S. there has been little temperature change in the past 50 years, the time of rapidly increasing greenhouse gases – in fact, there was a slight cooling throughout much of the country.

The GISS US historical temperature reconstruction has changed substantially since this 1999 version shown in Figure 6.5. Neither NOAA nor NASA publish charts showing the actual measured temperatures; all have been adjusted in some way.

Figure 6.5 NASA mean annual temperature anomaly

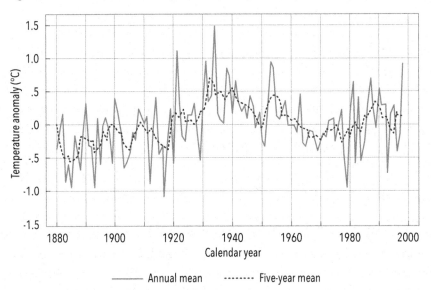

Source: Adapted from Hansen et al. 1999, http://www.giss.nasa.gov/research/briefs/hansen_07/fig1x.gif.

Global implications

In the same way that GISS adjusts temperatures from the US, it also adjusts temperatures from other parts of the world. This also results in record hot months, and years, for other places and the entire globe. For example, GISS recently published the following comment at its website:

> In 136 years of modern record-keeping, July 2016 was the warmest July according to a monthly analysis of global temperatures by scientists at NASA's Goddard Institute for Space Studies. Because the seasonal temperature cycle peaks in July, that means July 2016 also was warmer than any other month on record. July 2016 was a statistically small 0.1 degrees Celsius warmer than the previous record Julys in 2015, 2011, and 2009 (NASA 2016).

Just a few years earlier Dr Hansen had warned:

> The U.S. temperatures in the summer of 2012 are an example of a new trend of outlying seasonal extremes that are warmer than the hottest seasonal temperatures of the mid-20th century … The climate dice are now loaded. Some seasons still will be cooler than the long-term average, but the perceptive person should notice that the frequency of unusually warm extremes is increasing. It is the extremes that have the most impact on people and other life on the planet (NASA 2012).

According to GISS scientists, the average temperature for the contiguous US for 2012 was 14.6 °C, which was 0.6 °C above the mid-twentieth-century baseline.

Considering the actual unadjusted temperature measurements, however, it is inconceivable that NOAA and NASA could conclude that the summer of 2012 was hotter than 1936. None of the indicators from the NOAA daily raw temperature data point to the summer of 2012 being anywhere near as hot as the summer of 1936.

Astronaut Walter Cunningham wrote in 2010, that:

> NASA (GISS) and NOAA are uniquely positioned to debunk the current hysteria over AGW (anthropogenic global warming) but, unfortunately, they too appear to be caught up in the politics of global warming. Allowing their science to be politicized could destroy their credibility.

It is fashionable to draw parallels between those who question anthropogenic global warming and those who deny the Moon landing. Yet many astronauts, who have indeed been to the Moon and back, question anthropogenic global warming. They are concerned by the unbridled advocacy from GISS – and this includes the relatively recent significant, though unjustified, changes to the historical temperature record. The changes deny the hot years of the 1930s, and falsely suggest that the historical temperature record is consistent with anthropogenic global warming theory.

7 Taking Melbourne's Temperature
Dr Tom Quirk

The city of Melbourne has one of the longest surface-temperature records for anywhere in Australia. The Australian Bureau of Meteorology (the Bureau) makes adjustments to the actual measurements in the development of the Australian Climate Observations Reference Network – Surface Air Temperature (ACORN-SAT) official record, which, in turn, is used to determine long-term climate trends. Their adjustments, however, are in the wrong direction if the purpose is to make the official record more homogenous, and to compensate for the effects of urbanisation. In this chapter, I argue that the unadjusted raw maximum summer temperature as measured at Melbourne gives the best indication of long-term temperature change. This is calculated as an increase of 0.3 °C per century for the period 1856 to 2015.

Melbourne's weather stations

Melbourne, on the south-east tip of the Australian continent, was settled in 1835, and in 1851 it became the capital of the independent colony of Victoria. The temperature measurement record extends from 1856 to the present day, and is one of Australia's longest. Temperatures were first recorded on Flagstaff Hill, then at the Melbourne Observatory when it was established in 1863 near the Botanical Gardens.

The city was the gateway to the gold rush of the 1850s. Consequently, by 1860 the colony of Victoria was able to afford its own observatory, which became the home of the Great Melbourne Telescope. This was the largest telescope in the Southern Hemisphere and second largest in the world. The astronomers were interested in looking at the stars and observing the movements of the planets, as well as measuring temperatures, rainfall and humidity, and making weather forecasts.

In 1902, the government astronomer and meteorologist Pietro Baracchi, who was my great-uncle, supported the move to have the observatory relieved of this duty, and instead the meteorological work, carried out by astronomers, was placed under Commonwealth control. On 1 January 1908, with the establishment of the Bureau of Meteorology, the weather station was moved to the north-eastern edge of the junction of Victoria and La Trobe Streets in the central business district (CBD), as shown in Figure 7.1. This was a move of about three kilometres from the old Melbourne Observatory.

In 1908, the La Trobe Street site had been on the fringe of the least developed part of the city. But by the 1990s there were fifteen-storey buildings less than 100 m from the site, as shown in Figure 7.1.

Measurements were taken at this site until it was closed in 2015 – and the Stevenson Screen removed. Since 2013, measurements have been made at Olympic Park – a site two kilometres south and back towards the original observatory site.

So, there have been four measurement sites (weather recording stations) in a city that grew from a population of about 100,000 in 1854 to nearly four million by 2010.

Changes in equipment and recording times

Until 1964, temperatures were recorded at midnight, and then the thermometers reset for the next 24 hours. From 1964 to 1996 the temperatures

Figure 7.1 Location of Melbourne's weather station, 1908 to 2014

This photograph, of the La Trobe Street site was taken in 2017 three years after the Stevenson Screen (insert) was removed. The picture of the Stevenson Screen is from a Google, DigitalGlobe, image taken at this same site in 2014. This site is referred to in the official Bureau catalogue as the Melbourne 'regional office' site, but this is misleading, as it's actually in the central business district – opposite a 15-storey high tower.

Source: Google, DigitalGlobe for insert, and Institute of Public Affairs for street scene.

were read and reset at 9 am and 3 pm. Until 1996, maximum temperatures were recorded using a mercury thermometer, while minimum temperatures were measured using an alcohol thermometer.

In 1986, an automatic weather station (AWS) was installed at the La Trobe Street site. It measured temperatures alongside the mercury thermometer until 1 November 1996, when the AWS became the primary instrument replacing the thermometers. AWSs sometimes have a quite different housing to the weather station they replace, but in the case of the La Trobe Street weather station it was also housed in the Stevenson Screen, as shown in Figure 7.1.

When temperatures are measured using an AWS, the temperature is derived from a temperature-sensitive electrical resistance, known as a thermistor. The great advantage in using an AWS is that the recorded temperatures can be remotely measured with wireless communication, and the values read at any time.

For Melbourne, it is possible to download from the Bureau website daily minimum and maximum temperatures, and also the temperatures as recorded every 30 minutes through the 24-hour day. Figure 7.2 shows the average 30-minute frequency readings for June and December 2012 from the Melbourne La Trobe Street AWS. The minimum temperature occurs in the early morning, at about daybreak, while the maximum temperature is reached in the mid-afternoon.

The different temperatures that occur at such very different times of the day are from an atmosphere that is in very different states. The minimum temperature is subject to anomalies in the surface boundary layer of the atmosphere, which reflects local influences. The maximum temperature is from a large spatial volume of air and the presence of solar radiation that results in convective mixing of the atmosphere.

Interestingly, the mean maximum and minimum temperatures, which are generally used to calculate long-term climatic trends, give a different value to that calculated from the 48 readings taken every 30 minutes (weighted mean). The difference for December, a summer month, is 0.77 +/- 0.22 °C; while for June, a winter month, the difference is 0.37 +/- 0.09 °C. As minimum and maximum temperature records are averaged, it could be concluded that the Melbourne mean temperature is now overestimated by some 0.5 °C.

Figure 7.2 Daily temperature profile Melbourne, June versus December

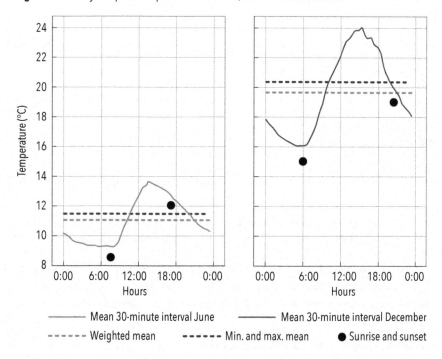

Temperatures measured at 30-minute intervals through a 24-hour day. The samples are for June (winter) and December (summer), 2012. The difference for (Tmin + Tmax)/2 − Tmean is 0.37 +/− 0.09 °C for June and 0.77 +/− 0.22 °C for December. Note the differing times for sunrise and sunset.

Source: Data downloaded from http://www.bom.gov.au/.

Summer versuses winter temperature maxima

Maximum temperatures are measured at a time of day when the atmosphere should be well mixed. This is especially the case in summer with the solar radiation present for some fourteen hours of daylight and some nine hours from sunrise, to reach the maximum temperature for the day.

Figure 7.3 Melbourne's maximum temperatures, summer versus winter

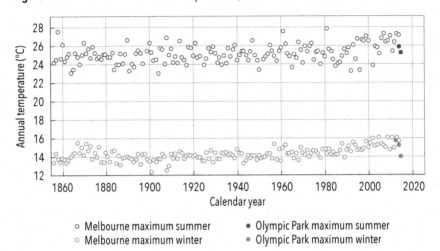

Legend	
○ Melbourne maximum summer	● Olympic Park maximum summer
○ Melbourne maximum winter	● Olympic Park maximum winter

Melbourne annual summer and winter maximum temperatures. The temperatures are based on the La Trobe Street site from 1908 to 2012, and then the Olympic Park measurements for 2013 to 2015 (including two summers and three winters). The Olympic Park site is not affected by screening of summer southerly winds.

Source: Data downloaded from http://www.bom.gov.au/.

Maximum summer and winter temperatures for Melbourne (including Olympic Park) are shown in Figure 7.3.

The summer maximum temperature shows a rise of 0.03+/– 0.02 °C per decade from 1856 to 1996 until the AWS was installed and a 15-storey new building constructed at 33 La Trobe Street. This is equivalent to a temperature increase of 0.3 °C per century. The scatter for the summer temperature after removing the trend is a standard deviation of 1 °C, and shows a random pattern of variations.

The maximum summer temperatures show no statistically significant increase.

What happened in 1996 to cause the temperature rise at La Trobe Street?

Daily temperatures can be averaged over a year, with the annual average raw minimum and maximum temperatures for Melbourne shown in Figure 7.4. The record covers Flagstaff Hill from 1856 to 1860, the Observatory from 1861 to 1908, and the Melbourne La Trobe Street site until its closure in 2015. The measurements from the new site in Olympic Park are *not* shown. The maximum annual temperature has a scatter of 0.50 °C but there is an apparent step-up in the temperature in 1996.

There is a classic statistical test called the Chow Break Test, which asks a simple question of whether a single time series is better considered as two independent series, that is broken into two straight lines, rather than one straight line. Considering Melbourne's maximum temperatures, the answer is that two straight lines represent a better fit to the record than one straight line, with 98% certainty, as shown in Figure 7.4. The temperature rise from 1856 to 1996 is 0.03 +/- 0.02 °C per decade, while the break is a difference of 0.7 +/- 0.2 °C.

The break in 1996 is likely the result of the construction of very tall buildings across the road from the La Trobe Street weather station that blocked cooling winds.

Winds significantly affect Melbourne weather, with the following description from the eMelbourne website (n.d.) relevant to understanding the break in the temperature series in 1996:

> The wind varies from day to night and from season to season. Wind speed is usually lowest during the night and early hours of the morning before sunrise. It increases during the day as heating of the earth's surface induces turbulence in the wind stream. Wind also varies, with much localised effects of some weather phenomena such as showers and thunderstorms. Examples of the diurnal variation are the sea breeze, which brings relief on many hot days, and the valley or katabatic breeze, which brings cold air from inland Victoria

Figure 7.4 Melbourne's annual maximum and minimum temperatures, 1856 to 2014

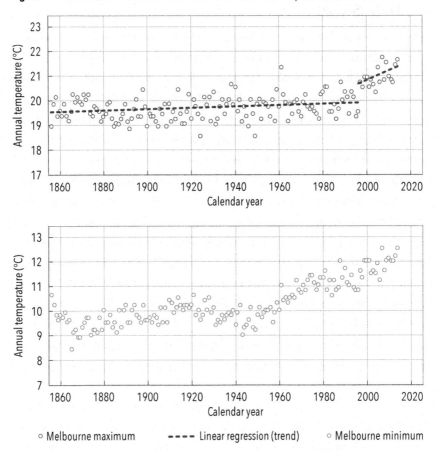

○ Melbourne maximum   **- - - -** Linear regression (trend)   ○ Melbourne minimum

Annual average minimum and maximum temperatures for Melbourne. For the maximum temperature, the two straight lines represent a better fit than one straight line with 98% certainty. The minimum temperature shows a more complicated pattern.

Source: Data downloaded from http://www.bom.gov.au/.

down valleys during the night and early morning towards Melbourne. These breezes are responsible for winds being more often from the north during winter, particularly during the morning. They are also responsible for winds being more often from the south during summer, particularly during the afternoon. This is in spite of the predominant wind stream being westerly in origin. There is a marked tendency for the very windy days to occur during the late winter and early spring months. Melbourne's strongest wind gust on record is 120 km/hr on 3 September 1982.

The key point to note, is that the maximum temperature break is coincident with the screening of the La Trobe weather station from summer southerly winds by the construction of the very tall City Gate building at 33 La Trobe Street, as shown in Figure 7.1. This is a fifteen-storey apartment tower completed in 1997, and to its east is a further apartment building completed in 1998.

So, the likely explanation for the 1996 increase is the screening of the site by the new buildings, but this does not exclude a contribution from the installation of an AWS. The possible impact of the AWS could be determined if the Bureau made the data from the AWS, as recorded from 1986, publicly available; at the time of writing only the AWS data from 1 November 1996 is available at the Bureau's website.

In 2014, the temperature at the new recording site of Olympic Park was 20.9 °C, while it was 21.7 °C at the La Trobe Street site. This is a difference of 0.8 °C – the same as the 1996 temperature increase at the La Trobe Street site.

The urban heat island effect

Cities are significantly warmer than surrounding areas due to what is known as the urban heat island (UHI) effect. Buildings, for example the wind-blocking new building at 33 La Trobe Street, can cause a local anomaly in the surface boundary layer of the atmosphere. For centuries, farmers have used hedgerows to shelter their orchards and crops from wind, and thus change their own local near-surface temperature. In a study of modifications to orchard climates in New Zealand, McAneney *et al.* (1990), showed that screening could increase the maximum temperature by 1 °C for a 10 m high shelter. In the humid coastal region that was the focus of their study, minimum temperatures were unaffected by the screening.

In order to calculate the UHI effect on Melbourne's temperatures, they can be compared with temperatures at Laverton, about 20 km to the south-west of the city centre. Temperatures have only been recorded at Laverton since 1944.

The raw maximum temperatures and the trends of maximum temperature in Melbourne are similar to that at Laverton, as shown in Figure 7.5. The two-temperature series are strongly correlated with a correlation coefficient of 85%. This can be seen in Figure 7.5, where the two records reflect the same temperature variations. However, the raw minimum temperatures as measured at Melbourne show a greater increase compared to Laverton from 1944 to the present. The increase at Melbourne, of about 2 °C compared to an increase of about 1 °C at Laverton, is most likely an indication of the much larger UHI effect in Melbourne.

The comparative increases, and the annual, summer and winter Melbourne–Laverton differences for the annual temperatures, are shown in Table 7.1. The minimum differences show that the UHI effect is detectable throughout the year. However, the maximum temperature shows a significant difference between summer and winter. The maximum temperature winter increase of 0.08 +/– 0.02 °C per decade is consistent with the winter maximum increase for the Melbourne measurement of 0.14 +/– 0.04 °C per decade.

Clearly there is a UHI effect in Melbourne, most clearly detectable in the minimum temperature measurements, of 0.2 °C per decade or 1 °C over a period of 50 years. That amounts to 2 °C per century.

Figure 7.5 Maximum and minimum temperatures, Melbourne versus Laverton

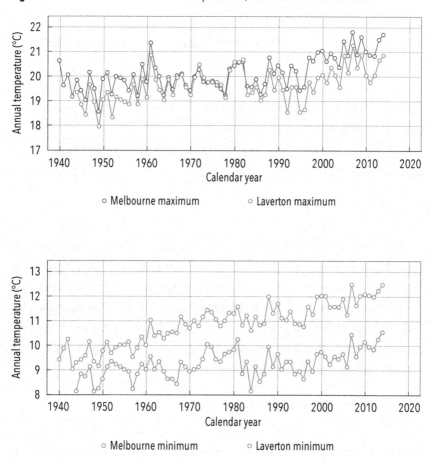

Annual average minimum and maximum temperatures for Melbourne and Laverton.

Source: Data downloaded from http://www.bom.gov.au/.

We are often told that global warming is occurring at about 1 °C per century and that this is something to be concerned about. Yet it would appear that Melbourne temperature minima have warmed by almost double this amount.

Table 7.1 Temperature increases from 1944 to 2014

Annual values	Minimum temperature increase °C per decade	Maximum temperature increase °C per decade
Melbourne	0.36 ± 0.02	0.22 ± 0.03
Laverton	0.15 ± 0.03	0.17 ± 0.03
Melbourne – Laverton	0.205 ± 0.016	0.053 ± 0.020
Seasonal differences for Melbourne – Laverton		
Summer (Dec, Jan, Feb)	0.206 ± 0.018	0.007 ± 0.030
Winter (Jun, Jul, Aug)	0.167 ± 0.021	0.080 ± 0.020

Source: Data downloaded from http://www.bom.gov.au/.

An increase is also detectable in the maximum winter temperature at Melbourne, but is not seen in the summer maximum to a significant extent. The behaviour of the maximum temperature perhaps shows that there is a common spatial volume of the atmosphere covering Laverton and Melbourne, which extends over at least 20 km. This would indicate that temperature maxima are not affected by urbanisation to the same extent as temperature minima.

The Bureau's adjustments

The Bureau makes changes to the temperatures as measured at the La Trobe Street weather station, and then incorporates these 'adjusted' values into the ACORN-SAT dataset. Values in this ACORN-SAT dataset are used to report climate variability and change by the Australian government, by CSIRO, and also by university researchers. Adjustments are made as step-changes, which are promulgated backwards in time, as shown in Table 7.2.

Two points need to be made. First, some environmental changes such as incremental warming from the UHI effect cannot be dealt with

Table 7.2 ACORN-SAT adjustments to Melbourne raw temperature measurements

Time series	Time of adjustment*	Reason	Adjustment °C
Maximum	1 January 1990	Statistical	0.41
	1 January 1996	Statistical	0.50
Minimum	1 January 1964	Time of observation	0.54
	1 January 1929	Statistical	−0.40

*Note: Adjustments are made to all data before this date.

by a step adjustment. Second, some of the changes that occur are real changes at the measurement site and not 'artificial biases'. Furthermore, the temperature should not be adjusted to an averaged value unless the area subject to the change is known and some weighting procedure produces an average temperature for a larger surface area. This difficulty is illustrated by the comparison of the La Trobe and Olympic Park measurements with a 0.8 °C change for a distance of only 3 km.

Apparently, the values in Table 7.2 have been calculated using an algorithm that considered neighbouring sites. I have determined, based on the information available from the Bureau, that some of these 'neighbouring' sites are more than 200 km from Melbourne, with Mount Gambier at 370 km and Larapuna (Eddystone Point) at 458 km on the north-east coast of Tasmania. It is difficult to see how such distant sites could be used to sense changes in the local Melbourne environment.

Three of the adjustments are explained as statistical, as shown in Table 7.2. No accompanying explanation regarding actual physical changes in the local environment are given, such as the impact of a weather station moving from one location to another, or a change in the measurement method.

The adjustment in 1964 is listed as a change in the 'time of observation', when the reading time of the thermometers changed from midnight to

9 am. When the thermometers were read at midnight then the minimum and maximum would be for that day, while a shift to a 9 am reading would give the minimum temperature for the day of the reading, but the maximum for the previous day. There is a possibility of double counting temperatures, however, this would be irregular and cannot logically be a step change assigned to all years as applied by the Bureau to Melbourne's temperatures as measured at the La Trobe Street site.

The 'adjustments' made to obtain what is referred to as the 'homogenised' ACORN-SAT Melbourne data are in the wrong direction to remove UHI anomalies. For example, the Bureau adds 0.41 °C to all temperatures as recorded at the La Trobe Street site before 1990. So, the adjustments have the effect of increasing past temperatures, rather than lowering the more recent temperatures. Clearly these adjustments do *not* compensate for the urban warming, but rather increase Melbourne's mean temperature.

Further, there is no sign of the 1990 step-change in the measured Melbourne maximum temperature record, as shown in Figure 7.4. It is unclear what statistical test justifies the step-up in maximum temperatures in the official ACORN-SAT record, as shown in Figure 7.6.

Changes to the minimum temperatures as measured at the La Trobe Street site increase past temperatures from 1995 back to 1940, as shown in Figure 7.7. These adjustments somewhat negate the real UHI effect.

Best estimate of long-term temperature change at Melbourne

If Melbourne's temperature as measured at the weather stations is to be used as an indication of long-term global change, then the maximum temperature in summer would give the best indication. This is because local temperature anomalies, such as the UHI, are not significantly affecting the record – at least not until 1996 at the La Trobe Street site. This is perhaps because there is opportunity for adequate mixing of the

Figure 7.6 Changing maximum temperatures as measured at Melbourne

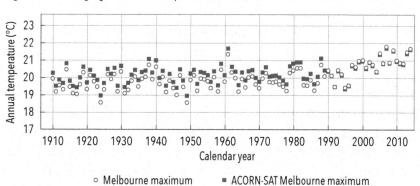

○ Melbourne maximum ■ ACORN-SAT Melbourne maximum

The official ACORN-SAT maximum temperatures for Melbourne are not the same as the maximum temperatures as measured at the La Trobe Street from 1908 through to 1990. The Bureau adds 0.41 °C to all maximum temperatures as recorded at the La Trobe Street weather station before 1 January 1990 back to 1908 – with annual mean maximum values as measured, and as calculated from the ACORN-SAT dataset, back to 1940 shown in this chart.

Source: Data downloaded from http://www.bom.gov.au/.

Figure 7.7 Changing minimum temperatures as measured at Melbourne

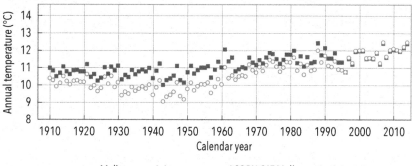

○ Melbourne minimum ■ ACORN-SAT Melbourne minimum

The official ACORN-SAT minimum temperatures for Melbourne are not the same as the minimum temperatures as measured at La Trobe Street from 1908 through to 1996. The Bureau adds 0.50 °C to all minimum temperatures as recorded at the La Trobe Street weather station before 1 January 1996 back to 1908, and 0.54 °C to all minimum temperatures before 1 January 1964 – with the effect of this change back to 1940 shown in this chart.

Source: Data downloaded from http://www.bom.gov.au/.

atmosphere during the longer summer daylight hours even over a large city such as Melbourne.

I calculate a rise in the summer maximum temperature at Melbourne of 0.03 +/− 0.02 °C per decade for the period from 1856 to 2015, which is equivalent to a rise of 0.3 °C per century. This rate of change is based on the Flagstaff Hill, Botanical Gardens, and La Trobe Street site data until 1996. I use the data from Olympic Park for the period 1996 to 2015.

This trend of 0.3 °C per century compares favourably with trends at nearby sites unaffected by UHI. For example, in a study of maximum temperatures at Cape Otway and Wilsons Promontory lighthouses, Jennifer Marohasy and John Abbot also found an overall annual change of 0.3 °C per century (Marohasy and Abbot 2015).

8 Mysterious Revisions to Australia's Long Hot History

Joanne Nova

Things are being done to Australia's national climate records that should never be done.

Something unnerving is going on in our datasets. A process called 'homogenisation' combined with other 'adjustments' creates improbable, almost comic effects – if only it were funny. For example, if the latest Bureau of Meteorology (the Bureau) figures are correct, the hottest day *ever* in the modern history of Australia did not occur in the baking arid zones of Oodnadatta or Marble Bar, but in Albany, in cold, far-south Western Australia (Nova 2014a).

Then there is the 'New Year's Eve' effect, where, if the Bureau corrections are right, each year thermometers that were over-estimating temperatures on December 31 suddenly start to underestimate them, and the whip-like corrections change by almost 2 °C from one day to the next. (So much for that tenth of a degree of accuracy.) Who knew that thermometers get hangovers?

The same thermometers, the Bureau would have us believe, develop a different error on the first day of the next month, February 1, and then swing the other way come March 1. What drove the Bureau to make these bizarre corrections? There is no wall, or tree, or shade, or breeze that generates errors according to alternate calendar months.

Another warning sign that things are amiss are the online Bureau temperature maps that feature almost supernatural detail. These maps display temperatures in the eastern Gibson Desert during World War I, even though there would be no thermometers sited there for another 40 years. Spookily, the temperatures shown for the 14 May 1914, are exactly the same as those shown for the 14 May 1915, and again on the same day for 1916, 1917, 1918, and on and on. Pick any random day – same 'groundhog day' effect. But hey, perhaps the weather did repeat each year?

In yet another example, the new highest-quality dataset suddenly contains nearly 1000 days where records say that the coldest point of the day was warmer than the hottest temperature that same day. Minima are not supposed to be warmer than maxima. These mistakes weren't generated by errors in old handwritten records but by high-speed gigabyte transformations. These simple errors were discovered by volunteers using standard basic quality control checks, yet two years after they were announced, the Bureau still hasn't fixed them.

In the mid-1990s electronic sensors were installed at sites across Australia. These can pick up very short bursts of heat, which means they can measure higher temperatures than the old mercury or liquid thermometers. The switch to the new thermometers appears to cause a 'step up' in temperatures in many places. The effect of the switch should be an easy question to resolve. The Bureau is running side-by-side comparisons, where measurements from the new electronic sensors are compared with measurements from the old thermometers. However, most of this data is not publicly available for comparison. Extraordinarily, this data is routinely deleted as a part of normal practice.

The Bureau has a $340 million budget, but seemingly can't afford an extra memory stick to save historic scientific data. These measurements from past years can never be re-recorded.

Citizen science fills the void

In the face of the relentless revision of our temperature history, I connected a team of volunteer citizen scientists, including Bill Johnston, a retired agronomist with experience measuring temperatures for New South Wales agriculture, Bob Fernley-Jones, a retired mechanical engineer, Ken Stewart, once a school principal, Lance Pidgeon, a free-lance radio technician, and Chris Gillham, a multimedia journalist. They and others in the team discovered that thermometers accurate to a tenth of a degree were being 'adjusted' by as much as two degrees. Original cooling trends were revised to warming trends. Often, the 'man-made' corrections were larger than the underlying trend.

These professional men and others comprised an independent audit group that teamed-up with Senator Cory Bernardi to ask for an audit of the Bureau's historical temperature dataset from the Australian National Audit Office (Nova 2011). The Bureau did not appear to welcome an audit. In the next twelve months, it dropped its long standing 'high-quality' dataset that the audit request referred to. In March 2012, the Bureau suddenly announced they had completed a major new revision of Australian historical data called the Australian Climate Observations Reference Network – Surface Air Temperature (ACORN-SAT). This had the effect of sidestepping the audit. Since the audit request applied to an old unused dataset, it could be officially ignored.

In 2014, when *The Australian* newspaper published reports of unex-plained major adjustments to the new ACORN-SAT dataset (Lloyd 2014a), another independent audit was proposed. In response, the same month, the Parliamentary Secretary to the Minister for the Environment set up a technical forum, which, in effect, ended up being a one-day event in 2015 which produced a report that looked only at statistical issues, not at the problems the volunteer team identified.

The forum resolved none of the important questions, but it did force the Bureau staff to admit that they would not be publishing a written method that anyone outside the Bureau could follow.

In effect, this practice of closed-shop secret methods has undertones of a kind of sacred guild of temperature artisans. It doesn't prove the many and varied 'adjustments' are wrong, but if they can't be repeated by independent outsiders, it isn't science (Nova 2015a).

The abandoned history of Australia

There is a forgotten history in old newspaper reports that gives us clues to historical weather patterns, and which would be wholly foreign to modern Australians. Carbon dioxide (CO_2) was lower in the 1800s than it is today, yet there are scores of references to temperatures of 125 °F 'in the shade' in our national newspaper archives. That is a dashboard-melting, scale-changing 52 °C.

Early explorers were trained to measure temperature. Charles Sturt recorded temperatures in the shade of 127 °F, 129 °F, and even 132 °F, reporting that the ground was so hot, if matches fell on the sand they would ignite spontaneously (Sturt 1849). In 1846, Sir Thomas Mitchell also recorded 129 °F (Mitchell 1846). He was afraid the thermometer

Figure 8.1 Newspaper headline from 1953

Source: Williams 1953.

would break as it 'only reached 132 °F'. In 1860, John McDouall Stuart's party measured 128 °F in the shade (Ewart 1861).

High temperatures were so common that in 1878 miners had a policy to 'knock off' work if the thermometer hit 112 °F (44.4 °C). There were no air conditioners then.

Despite these spikes of heat in the 1800s, by 1952 Australian scientists were discussing the cause of mysterious long cooling trends across a large part of the continent.

Few Australians would know that temperatures above 50 °C have been recorded in most states. In the future, when somewhere in Australia the temperature reaches more than 50 °C, it won't be the first time, but a repeat of past conditions.

History is being rewritten

Australia has some of the best historical data in the Southern Hemisphere, but there is little effort to accurately combine it with modern data. Since the fate of the world depends on understanding our climate, and we are spending billions to transform our economy, you might think we could afford to set up some wooden replicas of old historic screens near modern equipment and compare them. Surely it would be worth some effort to recreate the conditions under which historical datasets were recorded in order to connect them with today's datasets?

You might also think we could pay research teams to scour historical photographs, plans and documents for site changes *before* we adjust, or homogenise, the data. Instead, the Bureau ignores almost all data and other information from before 1910, and then does a poor-man's statistical fix – homogenisation – after that.

Some might think of old data as being primitive, but thermometers have been around for 400 years, and the technology was quite advanced by the late 1800s. The modern-style Stevenson Screen, which is used to house these thermometers, was rolled out across parts of Australia from 1889. At the time, scientists vigorously debated which type of screen was better. In 1887, Charles Todd was already running a side-by-side experiment that would eventually continue until 1948, a remarkable 61 years (Nicholls et al 1996). Those detailed experiments suggested there was about one degree difference between the two main types of screens; although, other researchers think it was less (Hughes 1995).

There are records from north Queensland where temperatures have been measured in Stevenson Screens from when they were first introduced in 1889; however, it was not until 1910 that *most* thermometers were housed in them. Cynically, though, one has to wonder, if the Federation Drought from around 1895 to 1902 had been incredibly cold instead of savagely hot, perhaps the Bureau may have been more interested in the early data recorded in these Stevenson Screens?

Today, the sometimes lower temperatures recorded in the same places that had these Stevenson Screens are hailed as 'unprecedented' records. Consider the case of Bourke. In 1908, a new thermometer with a Stevenson Screen was installed there. On 3 January 1909, a few months later, an observer recorded an extraordinary 125 °F, (51.7 °C). This record-breaking entry was handwritten and underlined as might be expected for a rare event. Decades later someone in the Bureau decided that this was an error because it was recorded on a Sunday when observers didn't normally work, and other towns in western New South Wales were 'only' hitting 45 °C. However, not far away in Brewarrina, newspapers listed the temperature there as 123 °F (50.6 °C). The question is, would someone who was a dedicated weather observer make an effort to come in to work on the hottest day ever recorded? It seems likely. It was probably the most exciting thing to happen in Bourke that day (Lloyd 2014b).

When the Bureau announces a record temperature in Bourke now, they don't mention this old, higher record.

How many Australians know that modern records often depend on adjusting the past – or in this case, deleting it (Trewin 1997; Marohasy 2017)?

There's a mismatch with historical records

Scientists a hundred years ago took their jobs seriously. Back in 1933, when the CSIRO was called the CSIR, the experts figured that they had 74 years of climate data and could put together a pretty useful collection. In the same vein, twenty years later, the Australian Bureau of Statistics (ABS) compiled year books of 84 sites from 1911–1940, with measurements taken from thermometers inside Stevenson Screens. Chris Gillham analysed that data, and estimates that nearly half the warming trend in Australia – about 0.4 °C of the 0.9 °C – comes from the adjustments. Furthermore, sites today are often artificially warmed by the urban heat island effect. Because of this, instead of adjusting down past temperatures, it's *more* likely that present temperatures should adjusted down to compare them properly with past ones.

It's fair to wonder, then, if the real warming trend in Australia might be a lot less than half a degree. In 1896, during the worst month of the Federation Drought, 437 people in the eastern states succumbed to the heat. Over the next 50 years Australia became slightly cooler and wetter. Areas from Bendigo to Alice Springs were 1–2 °C cooler, and had up to 20% more rainfall. These cooling trends covered about 2 million km^2 and were reported in a scientific paper (Deacon 1952), discussed by me in 2014 (Nova 2014b). A more recent study on the best sites in south-east Australia by Jennifer Marohasy and John Abbot shows the same pattern concluding that temperatures haven't changed much in 130 years (Marohasy & Abbot 2016).

Curiously, in 1953 the CSIRO officers thought the shift was either due to the sun (marked by changes in sunspots) or changes in the Antarctic polar ice cap (see Figure 8.1, Williams 1953; Nova 2014c).

Figure 8.2 Cooling trend reported in Australia to 1950

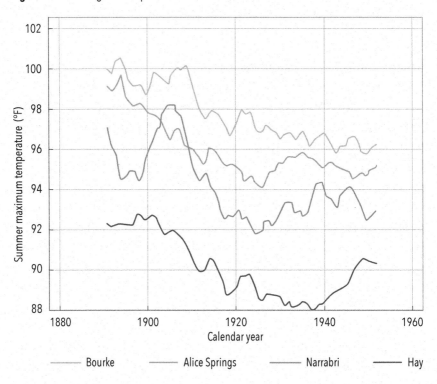

A study of summer maximum temperatures at four inland rural locations reported a general cooling trend.

Source: Adapted by permission from *Australian Journal of Physics* – Deacon, E.L., 'Climatic change in Australia since 1880', vol. 6, pp. 209-218, copyright 1952.

Sixty-two years later, not much has changed in the world of climate science – except the historical data, of course, which keeps being revised.

We know there is a problem

The Bureau claims that the adjustments have only a minor effect, but points to processed 'area-averaged' values to make this claim. Sceptics, meanwhile, compare individual stations on a site by site basis, and find large changes.

Area averaging is a technique to compensate for thermometers being spattered unevenly across the nation. Some lonely instruments represent thousands of square kilometres, while others represent much less. An average temperature estimated across the continent is not raw observation recorded by one instrument, and should not be called 'raw'. There are many different ways to calculate averages.

When Ken Stewart (2010) went through all the ACORN-SAT sites one by one, he found the minima across many of sites from 1910–1930 were reduced. The Bureau calls this 'neutral' but the warming trend increased by about 50%.

Signs the 'neutral' man-made adjustments are heating the nation
For the old high-quality (HQ) dataset, the Bureau said the adjustments were neutral. Dr David Jones, Head of Climate Monitoring and Prediction, National Climate Centre, Bureau of Meteorology, stated clearly that the adjustments made 'a near zero impact on the all Australian temperature' (Nova 2010). But, instead, Ken Stewart (2010) found they *increased* the trend by 40%.

When the HQ dataset was replaced with ACORN-SAT, the official description was carefully worded (Trewin 2012):

> There is an **approximate balance between positive and negative adjustments** for maximum temperature but a weak tendency towards a predominance of negative adjustments (54% compared with 46% positive) for minimum temperature.

The *number* of positive versus negative adjustments is not what matters. What matters is the *change* to the trend – which is increased by cooling adjustments to the oldest data, and warming adjustments to recent ones. The adjustments that increase the trend are larger. It's not balanced at all.

Figure 8.3 Locations with large adjustments based on temperatures at other sites

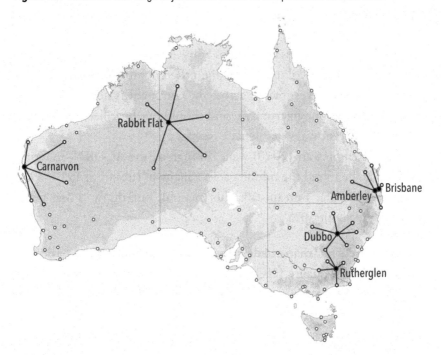

Adjustments to temperatures at six sites - Carnarvon, Rabbit Flat, Brisbane, Amberley, Dubbo and Rutherglen - increased trends by 2 °C. The adjustments are made relative to other 'nearby' sites as shown in the map.

Source: Adapted from Stewart 2014a.

When Ken Stewart compared the raw data on all 104 sites to the adjusted data, he found that the adjustments to the minimum temperatures increased the trend by 44.6% (Nova 2014d, Stewart 2014b). In some cases, temperatures are adjusted by as much as 2 °C (Stewart 2014a). These are supposedly our best stations – all temperatures were recorded after 1910 – and the trend we are looking for is about one degree a century, yet the adjustments in some cases are twice as large as the trend the Bureau is 'finding'. Also, some of the stations with large adjustments are in remote areas where their data is used to estimate temperatures over tens of thousands of square kilometres. Places like Rabbit Flat and Carnarvon are very influential in area-weighted estimates.

And the adjustments make little sense.

In the case of Carnarvon, the raw trend originally showed *almost no change* over a century (just 0.2 °C), similar to almost all its neighbours. However, after adjustments, the ACORN-SAT trend for Carnarvon finds it warming at the rate of 2.2 °C per century. After being 'corrected', Carnarvon is less like its neighbours than it was before.

The pattern is similar at all six sites as shown in Figure 8.3. It's difficult to see how this could be justified.

The Twilight zone of climate data

The 'hottest day ever' in Australia that nobody knows about

Here is a story of what must be a simple error that rewrites a key national record, and should have been found with the most basic of checks.

Chris Gillham discovered that the 'hottest ever day' in Australia according to the ACORN-SAT temperature set was not 50.7 °C, at Oodnadatta on 2 January 1960, which is written into history books. Instead it was the 8 February 1933 in, surprisingly, Albany, on the cool southern coast of Western Australia. This new record was created by the Bureau 79 years later through their new ACORN revision, which was announced in March 2012. This was a correction of a phenomenal 7 °C.

Ponder that even the people who lived in Albany that day in 1933 didn't know it was a record, and wouldn't find out for nearly 80 years until the Bureau 'determined' it wasn't 44 °C but 51 °C.

Errors like this tell us a lot about quality control at the Bureau. Surely 'the hottest ever' number is the simplest form of cross-checking – a ten-second search – but the Bureau hasn't done it and two years after this error was reported, the Bureau still hasn't fixed it (Nova 2014a). Is this world's best standard?

How many other errors lie undiscovered? How many days were invisibly warmed by 7 °C but are not so easy to find? This one change lifted the whole month's average.

The spooky climate in 1911: groundhog day

Lance Pidgeon spotted a strange thing that happened in central Australia back in World War I – each year, the weather repeated like some kind of groundhog day (Nova 2015b). Back then, in the eastern Gibson Desert, where the Northern Territory, South Australia and Western Australia meet in the middle of nowhere, where there were no thermometers, the Bureau can provide glorious daily weather maps for any day. But if you look closely, you will notice that each year's weather was just like every other year. It kept being repeated – the same wiggles, the same temperatures, and it wasn't just on the 14th of the month (as shown in Figure 8.4), but nearly every day. Pick a day, any day, spin the wheel, and watch the numbers come up the same. It wasn't just groundhog day, it was groundhog year. According to the Bureau, the same weather played out in 1916, then 1917, and so on, over and over. Some days repeat until 1956, when the magical repeating weather pattern was finally broken.

Does it matter? Given that the Bureau talks about area-averaged climates, yes, it does. We have to guess what the temperature was across most of central Australia, which means there's a lot of room to reinterpret the data.

New Year's Eve changes Perth thermometers by nearly 2 °C

Bob Fernley-Jones plotted the difference between the raw and adjusted data and was quite astounded to find a mysterious repeating square wave year after year, with the biggest shift in correction happening each

Figure 8.4 Weather pattern repeats until 1956

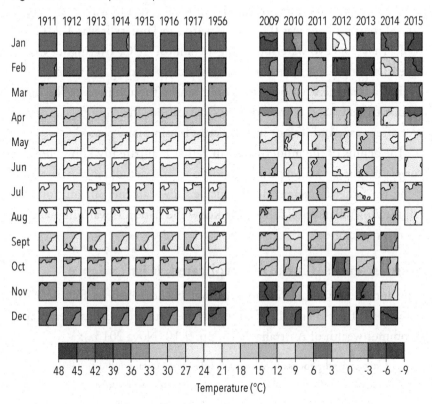

At this one zone in the Gibson Desert weather patterns repeated year after year in the early 1900s. Normal weather patterns vary (right-hand side).

Source: Australian Bureau of Meteorology website (http://www.bom.gov.au/jsp/awap/temp/index.jsp) and reprinted under Creative Commons (CC) Attribution 3.0 Licence.

New Year Eve. Apparently step corrections were regularly made according to the calendar month. (Like a kind of astrology for thermometers? As if a thermometer cared whether it was 31 December, or 1 January, or 1 February?) In Perth, these steps occurred through the whole record from 1910 onwards.

So, according to the Bureau's corrections, all the weather reports I listened to for years were wrong, day after day. In December, the Perth official thermometer was 0.8 °C too warm, in January it was 1.2 °C too cool. That's an eye-watering swing of almost 2 °C, overnight. Come 1 February, the same thermometer was back to reading too warm.

Try to imagine what situation could cause a thermometer to flip like a bipolar casualty, and require post hoc corrections in a 'monthly' pattern, year after year. And it's not just in Perth, but in many places.

The genius of the modern Bureau is that it apparently knows how to correct for these awful monthly conditions, and decades in the past. The thing is, if thermometers are this fickle, how can the Bureau honestly tell the Australian public that they know the nation is warming, and to a tenth of a degree?

Secret science: If it can't be replicated, it isn't science

The Bureau admits its methods cannot be described in full. In their own words their methods of adjusting the data are, with my emphasis (Commonwealth of Australia 2015, pp. 9-10; Nova 2015a):

> ... a **supervised process** in which the roles of metadata and other information required some level of expertise and **operator intervention.**
>
> ... several **choices** within the adjustment process remain a **matter of expert judgment** ...

Each station is adjusted using secret (but 'world-class') techniques, a bit like divining but without the long history.

Climate data is being destroyed

Researcher Bill Johnston wanted to compare datasets from electronic sensors to those of the old thermometers that the Bureau was running alongside. He wrote asking for six months of data from Sydney and Canberra Airports and in letters in reply from the Bureau he was told it would cost him $460 (Rea, Anthony 2014, pers. comm. reference: 30/5838, 23 December). Worse, that was only for the Sydney records. The Canberra ones had gone. The Bureau admitted that 'in accordance with records management practices' the field books for Canberra Airport are being 'disposed of' twelve months after the observations are taken. But that was better than nothing. The more recent Canberra Airport records didn't even *have* field books to be destroyed.

If the Tax Office took up the Bureau's data management plan, Aunt Kate's 1924 income would still be changing.

I can't delete my tax receipts, though, as far as I know, they're not being used to transform the economy.

For what it's worth, the $460 data fee was helpfully reduced to $230 after a lengthy appeal. In reality, the four-page assessment letter cost taxpayers more than that to produce, but this fee is a barrier preventing many taxpayers from analysing the data. Is that the point?

The Bureau has a budget of $340 million a year (Australian Government 2014); how much does it cost to store a text file?

The Future

Australia needs quality temperature sensors at quality sites, starting 100 years ago, although next week would be better than nothing. And after data is collected, we need an independent team to manage and store it, who are not the same people publishing climate papers and lobbying for different energy systems.

We audit banks, companies, government departments, energy flows, and projects, but we don't officially audit science. Whenever big money is involved we assume things need to be checked. When it's just the planet at stake, who cares?

The auditors need to be outside the climate science industry and outside academia. I suggest the job be given to independent hard scientists and engineers. It's only temperature trends fergoodnesssake.

9 The Homogenisation of Rutherglen
Dr Jennifer Marohasy

When climate scientists from Australia's Bureau of Meteorology (the Bureau) issue media releases at the beginning of each year telling us that last year was the hottest on record, most of us understand that to mean that they have simply added up the figures. We assume that last year *really* was the hottest. Very few of us are aware that, over time, the combination of weather stations used to calculate the average temperature for Australia has changed, or that individual temperature series are remodelled. In this chapter, I focus on the issue of remodelling. It has a technical name in climate science: it is called homogenisation. Homogenisation is justified on the basis that corrections must be made for non-climatic variables.

When a weather station is moved – for example, from a post office to an airport – this can cause discontinuities in the time series. However, climate scientists will even remodel (homogenise) perfect temperature time series: series without discontinuities, series that have been recorded from the same open field using standard equipment for more than 100 years. This is the case with the temperature series from Rutherglen – where temperatures have been recorded at the agricultural research station in northern Victoria since November 1912 – and in a Stevenson screen. The location and environment of the Rutherglen weather station is shown in Figure 9.1.

Figure 9.1 Location of the temperature recording equipment at Rutherglen

The white circle marks the location of the Rutherglen weather station, with the associated image showing a standard Stevenson screen. Temperatures at Rutherglen have always been measured in a standard screen.

Source: Map data from Google, DigitalGlobe; image of Stevenson screen from Bureau website (http://www.bom.gov.au/climate/cdo/about/airtemp-measure.shtml, viewed 14 May 2017) and reprinted under Creative Commons (CC) Attribution 3.0 Licence.

Rutherglen is part of the official Australian Climate Observations Reference Network – Surface Air Temperature (ACORN-SAT). The ACORN-SAT catalogue (Bureau of Meteorology 2012) clearly states, 'There have been no documented site moves during the site's history'.

Postmodern science and homogenisation

Like official (ACORN-SAT) temperature records, commercial milk is also homogenised. With respect to milk, the process is defined as the breaking down of fat globules so that they stay uniformly suspended in the liquid; it is about creating this uniformity. Similarly, in climate science, homogenisation allows scientists to remodel temperature data so there is a consistent story: a story that is 'closer to the heart's desire'. I have borrowed these few words – closer to the heart's desire – from the *Rubaiyat of Omar Khayyam*. The famous stanza reads:

> *Ah, Love! could thou and I with Fate conspire*
> *To grasp this sorry Scheme of Things entire!*
> *Would not we shatter it to bits – and then*
> *Re-mould it nearer to the Heart's Desire!*

Omar used the word 're-mould', whereas climate scientists say they are 'improving' the data. The end result, though, is the same; something has been changed.

While raw data was once considered sacrosanct, it is clear from the remodelling of Rutherglen's temperatures that the Bureau have rejected Enlightenment values, and embraced a postmodern philosophy – where raw data can be changed so that it accords with a much-loved theory.

While I use the homogenisation of Rutherglen's temperatures to illustrate my point in this chapter, the problem is more widespread, as explained by Professor Aynsley Kellow from the University of Tasmania in *Science and Public Policy: The Virtuous Corruption of Virtual Environmental Science* (2007).

The homogenisation technique used by the Bureau is also used by key institutions in the USA and the UK. Indeed, Gavin Schmidt – the director of the Goddard Institute for Space Studies (GISS), which is part of NASA, and the custodian of a key international dataset used by the United Nations' (UN) Intergovernmental Panel on Climate Change

(IPCC) – has stressed that when working with raw temperature data no one changes the trend directly, rather 'procedures correct' for perceived inhomogeneity based on a need for continuity with nearby stations. In the case of Rutherglen, a cooling trend of 0.3 °C per century in the observational data is changed to statistically significant warming of 1.6 °C per century – ignoring trends at nearby stations (Marohasy et al., 2014).

The actual measured temperatures from Rutherglen, un-homogenised

An historical temperature record is simply a sequence of data points, of successive measurements. Considering the Rutherglen temperature record, the Bureau began recording daily maximum and minimum temperatures in November 1912. These two series, the maxima and minima, can be combined to produce an average temperature for different time intervals, such as weekly, monthly, and also annual mean temperatures. When the annual mean temperatures as recorded at Rutherglen are charted (see Figure 9.2), we can get a visual understanding of the pattern of temperature change at Rutherglen.

The mean annual temperatures as measured at Rutherglen vary considerably between years. They show no overall warming trend, with the hottest years in this record being 1914 and 2007. These were both drought years, and many studies have demonstrated that low rainfall, causing drought in inland Australia, often results in higher air temperatures.

When the maximum temperatures for the hottest season in the year, summer, are combined and charted from the beginning of the record through to include the summer of 2015–2016, it is clear that the hottest summer on record was in 1938–1939 (see Figure 9.3). That summer was during a drought period when there were bushfires of unprecedented ferocity. At Rutherglen, the summer of 1938–1939 was more than two degrees hotter than the previous summer, and more than three degrees

Figure 9.2 Mean annual temperatures as recorded at Rutherglen

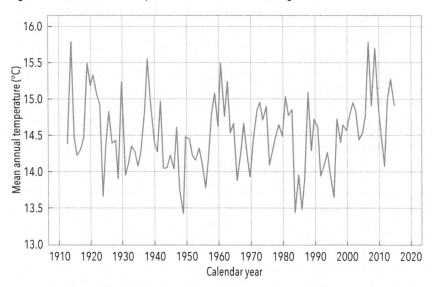

Source: Climate Data Online, Australian Bureau of Meteorology, September 2016, http://www.bom. gov.au/climate/data/.

hotter than the following summer of 1939–1940. This indicates that recent summers have not been particularly hot in historical terms, at least not at Rutherglen.

While we often hear that summers will get hotter with global warming, patterns of temperature change that are uniquely associated with the enhanced greenhouse effect concern the temperature minima – the night time temperatures. In particular, this popular theory predicts:

- greater warming of night-time temperatures than of daytime temperatures
- greater warming in winter compared with summer.

In other words, global warming theory predicts that minimum temperatures will warm more than maximum temperatures. However, when

Figure 9.3 Mean maximum summer temperatures as recorded at Rutherglen

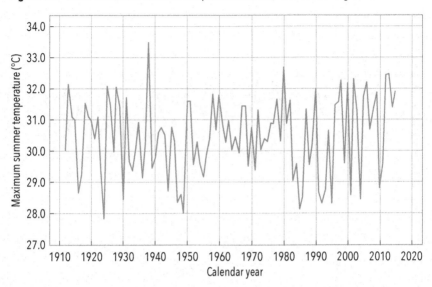

Summer is calculated from monthly data for December, January and February only. This chart includes the summers 1912-1913 to 2015-2016.

Source: Climate Data Online, Australian Bureau of Meteorology, September 2016, http://www.bom. gov.au/climate/data/.

we consider minima as measured at Rutherglen, and also three nearby locations, there is a cooling trend, as in Figure 9.4. The cooling trend is statistically significant at Benalla, Deniliquin, and Echuca. This does not accord with global warming theory. Temperatures at these locations should show more warming in their minima than their maxima, but the reverse is true. In fact, the minimum temperatures show cooling.

The apparently anomalous temperature trend in this region may be a consequence of land-use change associated with water infrastructure development.

Rutherglen, Benalla, Deniliquin, and Echuca are part of a watershed catchment known as the Murray–Darling Basin, and are not far from the

Figure 9.4 Mean annual measured raw minimum temperatures

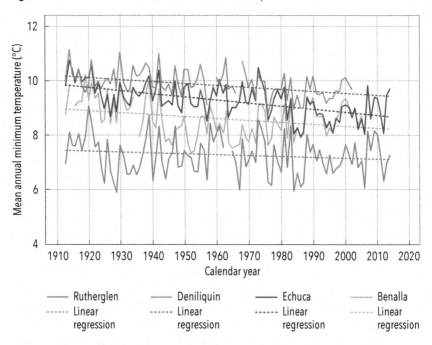

The chart shows the night-time temperatures as recorded at Benalla (1913–2005), Echuca (1913–2014), Deniliquin (1913–2002), and Rutherglen (1913–2014). The shorter records for Benalla and Deniliquin represent the longest complete records for these locations as measured at a single site.

Source: Climate Data Online, Australian Bureau of Meteorology, September 2016, http://www.bom. gov.au/climate/data/.

Murray River (see Figure 9.5). Since 1912, 800,000 hectares of irrigated agriculture, with water supplied from the Yarrawonga Reservoir, were developed in stages to the immediate west of Rutherglen. So this region, once associated with grazing of sheep, has become a hub for rice production with water pooled for long periods. This change in land use, in particular the presence of much more water in the environment, provides a possible physical mechanism to explain the long-term cooling trend.

Figure 9.5 Map of the Rutherglen region

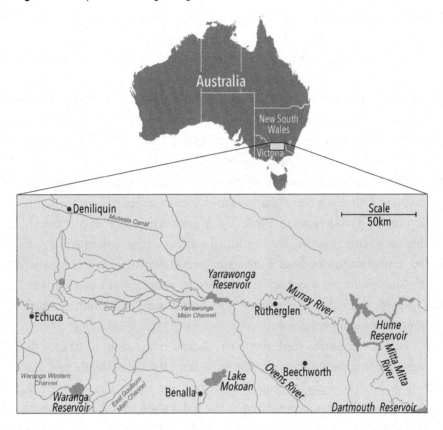

Locations of Rutherglen, Deniliquin, Echuca, and Benalla in south-east Australia, with the associated water infrastructure development shown in orange.

There are many such cooling temperature trends evident when raw measurements are plotted as recorded around Australia, and across the world. In the case of Rutherglen, the Bureau changes this cooling trend into warming through homogenisation – before incorporation into the official ACORN-SAT dataset.

How homogenisation works

It is possible to alter the trend in any time series by making specific adjustments to individual values, and then propagating these backwards to the beginning of the record. For example, the Bureau subtracted 0.61 °C from all temperature minima as measured at Rutherglen from 1 January 1974, back to the beginning of the record, as detailed in columns six, three and four of Table 9.1 respectively. They also subtract 0.72 °C from all temperature minima from 1 January 1966, back to the beginning of the record, also shown in Table 9.1.

These two adjustments have the effect of cooling the past relative to the present through two separate step-downs in temperatures: the reduction in 1974, and again in 1966. This is homogenisation as practised by climate scientists across the world. The effect – the consequence – on the temperature trends at Rutherglen are shown in Figure 9.6, where the blue circles represent annual minimum temperatures as recorded at Rutherglen, and the red squares represent annual minimum temperatures after homogenisation.

The adjustments cool the past. Linear regression of the actual measured temperatures is shown with the blue-coloured dashed line, which shows a slight cooling. In contrast, the linear regression of the homogenised values, represented by the red dashed line, shows warming. This demonstrates how the homogenisation of the Rutherglen temperature record has changed a slight cooling trend into a warming trend.

The five 'adjustments' listed in Table 9.1 are purported to correspond with discontinuities, also known as breakpoints, in the data.

In a two-page advisory issued by the Bureau of Meteorology (2014), it is stated that the total of five breaks, as listed in Table 9.1 (three in the temperature maxima, and two in the temperature minima), were detected using statistical methods when the data from Rutherglen was compared with 'surrounding stations (neighbours)'. The advisory also states that

Table 9.1 ACORN-SAT Rutherglen adjustment summary

Station name	Station number	Temperatures adjusted	Date (adjustment applied to all data prior to this date unless stated)	Cause	Impact of adjustment (°C)	Seasonal (if applicable)	Comparative stations	Notes
Rutherglen	82039	Min	1/1/1974	Statistical	−0.61		74034 82053 82002 72097 82100 74106 81049 81084 72023 82001	
Rutherglen	82039	Min	1/1/1966	Statistical	−0.72		82053 82002 82001 72150 74114 80015 74039 74062 74128 75032	
Rutherglen	82039	Max	1/1/1965	Statistical	−0.39		82053 72023 82001 82002 75028 75031 80043 72000 72150 80015	
Rutherglen	82039	Max	1/1/1950	Statistical	−0.62		82053 72023 82001 82002 75028 75031 80043 72000 72150 80015	
Rutherglen	82039	Max	1/1/1938	Statistical	−0.62		82053 72023 82001 82002 75028 75031 80043 72000 82016 80015	

Source: Bureau of Meteorology 2014.

Figure 9.6 Homogenisation of Rutherglen's minimum temperatures

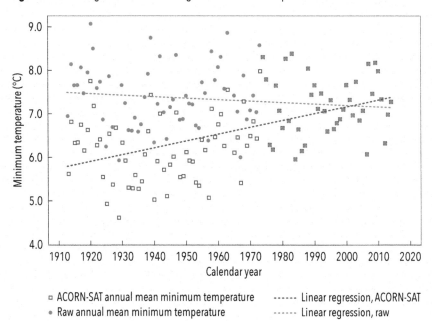

□ ACORN-SAT annual mean minimum temperature ------ Linear regression, ACORN-SAT
• Raw annual mean minimum temperature ------ Linear regression, raw

This chart shows the raw annual minimum values in blue, and the homogenised annual minimum values in red. They correspond after 1974, but diverge before 1974 because of the adjustments shown in Table 9.1.

Source: Adapted from Marohasy, J. 2016, 'Temperature change at Rutherglen in south-east Australia', *New Climate*, http://dx.doi.org/10.22221/nc.2016.001.

23 neighbouring stations were used 'at various times, combined with the use of documentary records', and that this 'reveals that there have been five significant breaks in the data'.

It is stated in fourteen different places in this advisory that 'statistical methods' were used to detect the breakpoints relative to the neighbouring stations. But at no time is the statistical method explained, nor are the breakpoints shown graphically.

One of the neighbouring weather stations listed in the Bureau's advisory is Beechworth. Beechworth is station number 82001, as listed in column 10 of Table 9.1. Beechworth is approximately 40 km south-east of Rutherglen. Minimum and maximum temperatures were recorded for this location from January 1908 until June 1986.

When monthly temperature minima from Beechworth are run through a control chart using standard commercially available Minitab software (version 17), we see that there is a very obvious breakpoint (discontinuity) in this temperature series occurring from 1976. This is shown in the top chart of Figure 9.7. This breakpoint corresponds with exceedance of the upper control limit (UCL) in the middle chart, see Figure 9.7.

Figure 9.7 Statistical assessment of monthly temperature data from Beechworth

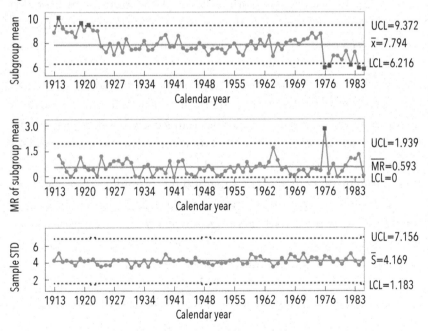

Source: I-MR-R/S control chart (Minitab version 17) showing annual mean temperature minima (top chart), moving range of annual mean (middle chart), and standard deviation of annual mean (bottom chart) for monthly temperature data as measured at Beechworth (1913–1985).

In fact, this temperature series could not be considered an accurate representation of temperatures at Beechworth – because the temperature measurements have been affected by a non-climatic event. This breakpoint, represented by the step change in the temperature series from 1976, actually corresponds with a documented site move at Beechworth (Marohasy 2016). This abrupt change in placement of the Stevenson screen appears to have caused the step-down in the annual minima (top chart), and a corresponding exceedance of the upper control limit for the moving range (middle chart). The sample standard deviation appears unaffected as shown in the lower chart (see Figure 9.7).

It would therefore be appropriate to make adjustments to the temperature series at Beechworth to account for this discontinuity associated with the weather station having been moved – at least this would be in accordance with Bureau policy. But, instead, the Bureau use the temperature series from Beechworth to homogenise temperatures at Rutherglen.

Yet when the minima temperature series for Rutherglen is run through a control chart, there are no breakpoints – there are no obvious discontinuities (see Figure 9.8). To be clear, we see that the mean annual minimum temperature (top chart) fluctuates within three standard deviations (defined by the upper and lower red lines) from the overall mean. The moving range of the subgroup mean (middle chart), and the sample standard deviation (bottom chart) are also generally in control for the period of the record. This suggests that *if* there had been any site moves or equipment changes at Rutherglen they have *not* significantly perturbed the historical minima temperature record.

Considering the control charts for both Beechworth (see Figure 9.7) and Rutherglen (see Figure 9.8), it would be absurd to change the 'in-control' temperature series at Rutherglen based on the 'out-of-control' series from Beechworth – yet this is exactly what the Bureau does.

Figure 9.8 Statistical assessment of monthly temperature data from Rutherglen

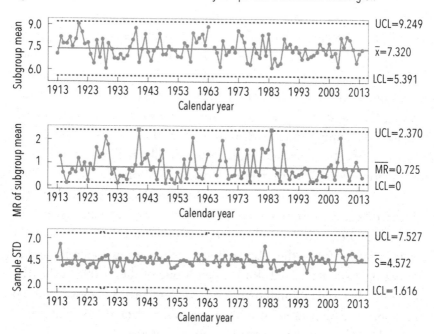

Source: I-MR-R/S control chart (Minitab version 17) showing annual mean temperature minima (top chart), moving range of annual mean (middle chart), and standard deviation of annual mean (bottom chart) for monthly temperature data as measured at Rutherglen (1913–2014).

Control charts are routinely used for quality control within many other disciplines that analyse time series data (Ryan 1989; Taylor 1991). The technique does have much potential application to climate science, and I have used it to find discontinuities, and correct the same, in temperature series from Cape Otway lighthouse (Marohasy & Abbot 2015). The Bureau *does not* use control charts to find, or correct, discontinuities. The Bureau uses *unique* algorithms and its own software, which are not publicly available – their results cannot be easily replicated and the breakpoints they report are not evident when the data is charted.

Political stonewalling

Concerned with claims that the Bureau was exaggerating estimates of global warming and justifying this through the dubious application of statistics, former Australian prime minister Tony Abbott's own department proposed setting up an investigation in late 2014. The then environment minister, Greg Hunt, acknowledged on Australian Broadcasting Corporation's *Lateline*, on 24 September 2015, that he 'killed' the idea, explaining to journalist Tony Jones that he was confident the Bureau used, 'hard science, hard data, literally millions of points of information through our satellite and local monitoring'.

The Bureau, in fact, *does not* rely on satellite data in the development of the official historical temperature record for Australia. ACORN-SAT is based entirely on surface temperature records from 112 weather stations, with 109 of these being homogenised using a technique that cannot be easily replicated. Rutherglen is one of these 109. In the case of Rutherglen, even though there are no documented site moves, and no breakpoints (discontinuities) in the temperature data, the Bureau nevertheless homogenises the minimum temperature series.

To conclude, minimum temperatures from this weather station have been remodelled by the Bureau so that a mild cooling trend becomes statistically significant warming in the official homogenised temperature reconstruction. This makes the temperature trend for Rutherglen more in accordance with the theory of anthropogenic global warming, but it is inconsistent with the reality.

10 Moving in Unison: Maximum Temperatures from Victoria, Australia

Dr Jennifer Marohasy & Dr Jaco Vlok

Climate change is often described by complicated statistics. In this chapter, we simply show all the maximum temperature series ever recorded by the Australian Bureau of Meteorology (the Bureau) for Victoria, Australia. We encourage you to make up your own mind about temperature trends by simply looking at this raw data. The individual temperature series show a very high degree of synchrony – they move in unison, suggesting they are an accurate recording of climate variability and change. But there is no long-term warming trend. There are, however, cycles of warming and cooling, with the warmest periods corresponding with times of drought.

The temperature series

Central to any analysis of or discussion about climate change and its causes is the historical temperature data. This record is often presented as a simple series of annual values – sometimes these are labelled 'observations' (for example, Lewis & Karoly 2014) even though they may actually represent a complex compilation of hundreds of remodelled series incorporating complex statistics (Brohan et al. 2006). The original measurements may have been recorded from radiosondes in weather balloons, from satellites, or from liquid-in-glass thermometers at weather stations. Very long temperature series, which may extend back several thousand years, are often derived from proxy records – such as tree-ring measurements.

For Victoria, in south-east Australia, there are continuous temperature measurements from thermometers, and more recently from temperature probes. There is no one continuous record for the entire state of Victoria, but rather nearly 300 individual temperature series of different lengths, representing measurements taken at weather stations across this region.

At these weather stations the maximum and minimum temperatures have been recorded each day. It is possible to download the daily values, and the monthly averages from the Climate Data Online (CDO) archive at the Bureau's website. Figure 10.1 shows the plotted maximum and minimum monthly temperatures as recorded at the Rutherglen research station, in northern Victoria. The series begins in November 1912, and is almost complete through to the end of 2016. The record is missing data from August 1928 (maximum temperature only), April 1963, and all of 1964.

The monthly values show a clear sinusoidal pattern with one cycle per year corresponding to the seasonal variability. Long-term variability becomes more apparent through the use of the twelve-month moving (or rolling) average, also shown in Figure 10.1 for both maxima and minima temperatures. Both maxima and minima are often combined and divided by two to show the mean temperature. However, sometimes, for example during drought, the temperature maxima and minima diverge (that is, the days are hotter and the nights are cooler). In this chapter, we focus on the temperature maxima, which are in effect a measure of afternoon temperatures[1] (the hottest time in a 24-hour cycle).

---

1   Temperature minima are also recorded, but these are not shown nor discussed in this chapter. They are shown and discussed in 'Analysis of Historical Temperature Data of Victoria' (Vlok 2017).

Figure 10.1 Monthly data from the Rutherglen research station

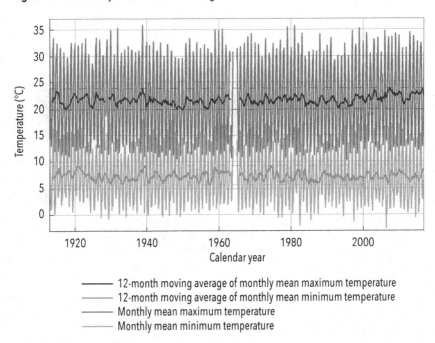

——— 12-month moving average of monthly mean maximum temperature
——— 12-month moving average of monthly mean minimum temperature
——— Monthly mean maximum temperature
——— Monthly mean minimum temperature

Source: Australian Bureau of Meteorology website at http://www.bom.gov.au/climate/data/.

Across Victoria, individual weather stations have recorded maximum temperatures that are consistently slightly warmer or cooler than each other, with the gradient generally consistent across the years. So, the individual series can be, more or less, neatly stacked one on each other, as shown in Figure 10.2 (top chart).

Figure 10.2 (top chart) shows only the temperature maxima, charted as twelve-month moving averages for every available maximum temperature series that can be downloaded from the Bureau's CDO webpage for Victoria. The mean of these values is shown as the black line in Figure 10.2 (top chart), which fluctuates at around 20 °C.

Figure 10.2 Maximum temperatures as recorded at Victorian weather stations

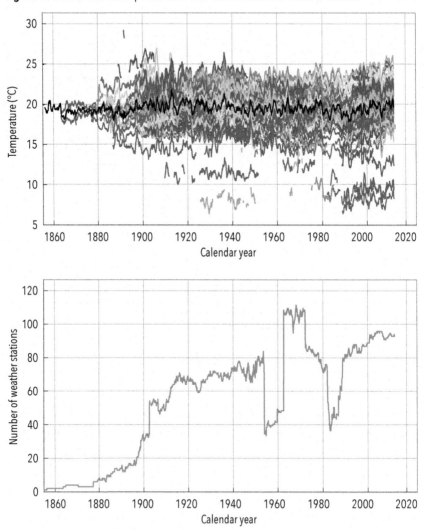

The top chart shows all the individual maximum temperature series – represented by different colours. The bottom chart shows that the number of weather stations recording maximum temperatures has changed, affecting the number of series used to measure climate variability and change.

Source: Australian Bureau of Meteorology website at http://www.bom.gov.au/climate/data/.

The number of weather stations recording maximum temperatures across Victoria has changed since 1855. When records began, there was a single weather station measuring temperatures in the centre of Melbourne. The number of stations increased gradually up until 1956, but then reduced dramatically for a time before climbing gradually back up to more than 90 (Figure 10.2 bottom chart).

Weather stations in the cooler alpine Eastern Highlands – near the ski resorts of Mount Buffalo, Mount Buller, Hotham Heights, Falls Creek, Mount Hotham, and Mount Baw Baw – have mean maximum temperatures that fluctuate around 10 °C, as shown in the bottom, right-hand corner of Figure 10.2 (top chart). Some of these stations operated before 1980, but were inactive for several years and then started recording again in the late 1980s and early 1990s. Others were only installed after 1980.

While there are continuous records available from 1855, the Bureau only begins the official temperature reconstruction for Victoria, and Australia, in 1910. This is ostensibly because some temperature recordings before this time were made in non-standard equipment (that is, the thermometers were not inside Stevenson Screens), and therefore the data may not be reliable. In this chapter, to reduce the potential for controversy, we only use the period from 1910 to 2016 to calculate temperature trends.

Cooling and warming cycles correspond with periods of drought and flood

The period 1910 to 2016 can be broken into shorter, approximately 35-year time frames, with the first of these periods shown in Figure 10.3. Displaying the series in 35-year blocks facilitates a more detailed visual assessment of temperature variability and change at the decadal scale.

The early record, from 1910 to 1945, indicates that mean maximum temperatures – as recorded from the approximately 60 weather stations operating at that time – oscillated in unison within a band of

Figure 10.3 Maximum temperatures 1910 to 1945

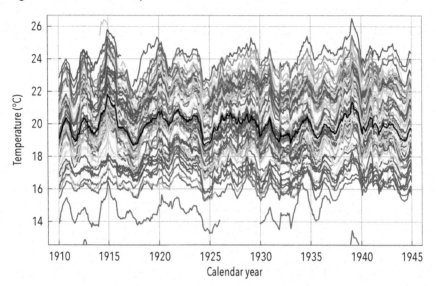

Trends for the periods 1945 to 1980, and 1980 to 2016 are shown and detailed in 'Analysis of Historical Temperature Data of Victoria' (Vlok 2017).

Source: Australian Bureau of Meteorology website at http://www.bom.gov.au/climate/data/.

approximately 14 °C to 26 °C, as shown in Figure 10.3. The temperature initially climbed to a peak in late 1914, before falling off dramatically. The rise and fall was of approximately two degrees over a period of just a few years.

This temperature spike corresponded to a period of below average rainfall, with the Murray River running dry in the autumn of 1914 – the first time this happened since the beginning of European settlement. There was another terrible drought from 1938 to 1940, which, again, presents as a period of higher temperatures in Figure 10.3; temperatures rose and then fell by approximately two degrees. These drought periods also corresponded with very low Southern Oscillation Indices of –14, and –16, respectively. The droughts broke with flooding rains, which then corresponds to the years of below average temperatures.

While it might seem reasonable to conclude from Figures 10.2 and 10.3 that temperatures are cycling up and down, rather than generally increasing across Victoria, this conclusion is not consistent with the annual climate statement from the Bureau. The Bureau reports a general warming trend consistent with the global increase in concentrations of carbon dioxide (CO_2) for Victoria, and Australia more generally.

Temperatures series used to calculate official statistics for Victoria

Since 1996, the Bureau has calculated and reported the annual change in temperatures for each state and for the entire nation – with this work initially funded by the National Greenhouse Advisory Committee (Torok & Nicholls 1996). Previously, it was more usual to report climate change in 30-year blocks, linked to shifts in the mean position of the subtropical high pressure belt (for example, Deacon 1953). The first dataset used to calculate the annual values is known as the Torok and Nicholls dataset, and included 27 stations for Victoria. This dataset was superseded by the Della-Marta et al. series (2004), which included 23 series and was used from 2002 to 2011.

Since 2012, the annual climate statements (Bureau of Meteorology 2016) have been based on the ACORN-SAT dataset, which combines just sixteen Victorian weather stations – according to the ACORN-SAT station catalogue. Some weather stations from other states also contribute to the Victorian value, because the individual values are placed on a grid network. So, Mount Gambier in South Australia and Cabramurra in New South Wales also contribute to the mean temperature as calculated for Victoria in 2016 (Trewin 2017, pers. comm. 18 January). However, before any of the values from these weather stations are used in the calculation of the official state and national annual averages, they are first homogenised and incorporated into ACORN-SAT (Trewin 2013).

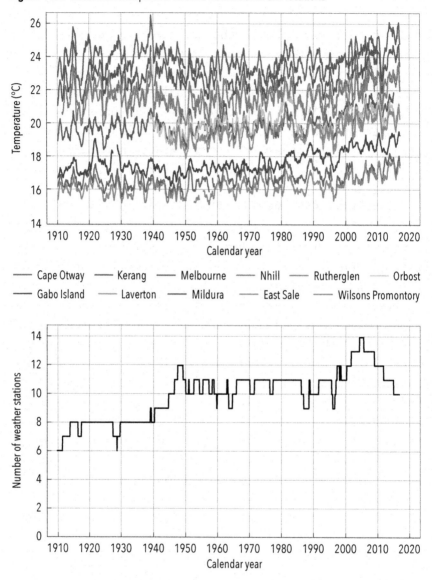

Figure 10.4 Maximum temperature series from ACORN-SAT locations

Legend: Cape Otway — Kerang — Melbourne — Nhill — Rutherglen — Orbost — Gabo Island — Laverton — Mildura — East Sale — Wilsons Promontory

The raw temperature series used in this chart are subsequently homogenised, before inclusion in the ACORN-SAT dataset. The ACORN-SAT dataset is used to generate the official temperature statistics for Victoria, and Australia. The bottom chart shows that the number of weather stations at ACORN-SAT locations has changed.

Source: Australian Bureau of Meteorology website at http://www.bom.gov.au/climate/data/.

The raw series from the sixteen Victorian weather stations, which are subsequently homogenised and incorporated into ACORN-SAT, are shown in Figure 10.4 (top chart). Temperature trends at several of these weather stations, including Melbourne, also include artificial local warming from the urban heat island (UHI) effect. These urban sites are not included in the Bureau's calculation of the area averages. The Melbourne regional office is represented by the red line in Figure 10.4, starting at about 19 °C.

While Melbourne is not used to calculate the trend for the annual climate statement, the Melbourne temperature series is used to homogenise series that are used to calculate the trend for the annual climate statement. For example, the temperature series for Cape Otway lighthouse is homogenised through the application of algorithms that make it more like the Melbourne series (Marohasy & Abbot 2015).

The Cape Otway and Wilsons Promontory lighthouses are the series with the lowest mean temperatures in Figure 10.4 (top chart), shown by the two bottom lines, in orange (Cape Otway) and turquoise (Wilsons Promontory). Clearly, the rate of warming at these lighthouses is not as pronounced as at Melbourne.

Mildura in the far north-west of Victoria has the highest mean maximum temperature, shown by the top dark blue line in Figure 10.4 (top chart). Inland locations are generally hotter, and are more affected by droughts and floods, and therefore temperatures fluctuate much more than at coastal locations.

Temperatures at Mildura were initially recorded at the post office, then a weather station was opened at the airport in 1946. For a couple of years these stations were both in operation, as shown in Figure 10.4 (top chart). Mildura is now a regional centre with a population of approximately 50,000.

There is only one long, continuous, inland temperature series from Victoria recorded from the same site that is not affected by a UHI. This is the series from the agricultural research station at Rutherglen. The maximum and minimum monthly values are shown in Figure 10.1, and also represented by the green line that averages 21.8 °C in Figure 10.4 (top chart).

Combining temperature series, and calculating temperature trends

Changes in temperature over time are typically characterised through linear regression. That is, when we hear that there has been a 1 °C increase in temperatures during the last 100 years, this typically refers to the slope of the regression line calculated by plotting the annual values. Obviously, the mean annual maximum temperatures used to calculate such a trend will vary with the geography of Victoria. For example, temperatures recorded at alpine ski resorts will, on average, be much cooler than temperatures from the Mallee country. This is evident in Figure 10.5 that shows the location of individual weather stations, and their average maximum temperature.

Since 1910, the number of weather stations in alpine areas, and around Melbourne in the south, has increased, while the number of sites in the relatively hot Mallee region of the north-west has decreased. This change in the mix of stations drags down the state-wide monthly average temperature, as shown by the black line in Figure 10.2. So, even if there has been no overall change in the climate, linear regression through the mean of all the stations shows an overall cooling trend.

This is the case when the linear regression of all 289 maximum temperature series is calculated, as shown in column 4 of Table 10.1. It suggests that for the period 1910 to 2016, there is an overall cooling trend in the temperature maxima for Victoria. However, this is mostly likely an artefact of the method used to combine the individual series.

It is impossible to replicate the method used by the Bureau to calculate the current ACORN-SAT values. This is not only because of the

Figure 10.5 Locations and average maximum temperature

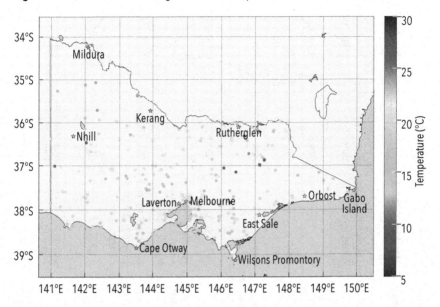

Only ACORN-SAT stations are named in this chart, but the individual dots represent every weather station for which data is available. The colour of the dot gives an indication of the average maximum temperature at that location for the period of the record.

Source: Australian Bureau of Meteorology website at http://www.bom.gov.au/climate/data/.

complexity of the homogenisation system used to remodel individual series (see chapter 9 for a discussion of the homogenisation of Rutherglen), but also because the maximum temperature anomalies for each state are based on the application of a complex weighting system to each of the individual homogenised series, specifically in response to a report to a government review of the ACORN-SAT dataset. The Bureau has acknowledged:

Temperature anomalies are based on daily and monthly gridded data with more than one station contributing towards values at each grid point.

Table 10.1 Annual trends for each dataset, and information on the method used to calculate the statistics

Reconstruction	ACORN-SAT, Victoria	Marohasy & Abbot I.	Marohasy & Abbot II.	All raw, Victoria	Raw ACORN-SAT, Victoria
Dataset	ACORN-SAT daily homogenized values, Bureau	Raw monthly values, Bureau	Raw monthly values, Bureau	Raw monthly values, Bureau	Raw monthly values, Bureau
Number of series used	16	5	5	289	16
Same series used to calculate every value	No	Yes	Yes	No	No
Type of data	Homogenised	Raw, & two series adjusted for equipment changes	Raw	Raw	Raw
Period	1910-2016	1887-2013	1910-2013	1910-2016	1910-2016
Area weighting	Yes – gridded system without specific set of weights	Yes – consistently 45:54:1 for coastal: inland: urban	Yes – consistently 45:54:1 for coastal: inland: urban	No	No
Geographic area	State of Victoria	Landmass 140° east longitude, 35° south latitude	Landmass 140° east longitude, 35° south latitude	State of Victoria	State of Victoria
Trend: annual maximum (degree Celsius per century)	+1.0 °C	+0.3 °C	+0.8 °C	-0.5 °C	0.7 °C
Visual representation	Red series in Figure 10.6	n/a	Orange series in Figure 10.6	Blue series in Figure 10.6	n/a

Source: Australian Bureau of Meteorology website at http://www.bom.gov.au/climate/data/ or http://www.bom.gov.au/climate/change/acorn-sat/#tabs=Data-and-networks.

Unlike simpler methods such as Thiessen polygons, there is no specific set of weights attached to these. The effective contributions change on a daily or monthly basis (Bureau of Meteorology 2015).

A much simpler and more transparent system was used in an alternative reappraisal of temperature trends for south-east Australia (Marohasy & Abbot 2016). In this study, Marohasy and Abbot used only five maximum temperature series to understand climate variability and change since 1887. These were: Melbourne; the two lighthouses with long records (Cape Otway and Wilsons Promontory); and the two inland locations with the longest and most complete records (Deniliquin and Echuca). Rutherglen was not included because this series only begins in 1912. (They wanted temperature series that began on, or before the Federation Drought of 1895–1902.) After considering the geography of the region (specifically, the area east of longitude 140, and south of latitude -35), some arbitrary decisions about relative weightings were made: in particular, that temperatures at Echuca and Deniliquin would be representative of 54% of the landmass of south-east Australia, the lighthouses 45%, and Melbourne 1%. (Melbourne only covers about 0.02% of the landmass of Victoria, however, other places in south-east Australia are also likely to be affected by UHIs.) These area weightings were applied, in order to develop a single temperature series for south-east Australia for the period from 1887 to 2013.

The overall linear trend for the Marohasy and Abbot series indicates a rate of warming of just 0.3 °C for the period from 1887 to 2013, as shown in Table 10.1. When the series is truncated to begin in 1910, the rate of warming more than doubles to 0.8 °C per century. This demonstrates that any trend is also sensitive to the start date of the series.

The Marohasy and Abbot (2016) reconstruction for south-east Australia shows an almost identical pattern of inter-annual variability to the single maximum temperature series from Rutherglen, as shown in

Figure 10.6. The Rutherglen series only begins in 1913 and has a rate of warming of 0.5 °C per century.

The ACORN-SAT series for Victoria has a rate of warming of 1 °C per century – and is shown by the red line in Figure 10.6.

The black line from Figure 10.2 (top chart), that is the mean of all the 289 series, is shown by the blue line in Figure 10.6. Relative to the other series, this reconstruction represented by the blue line in Figure 10.6 shows cooling from 1980, which, at least in part, is a consequence of the addition of the alpine series with no corresponding area weighting applied to off-set this.

Figure 10.6 Comparing temperature trends

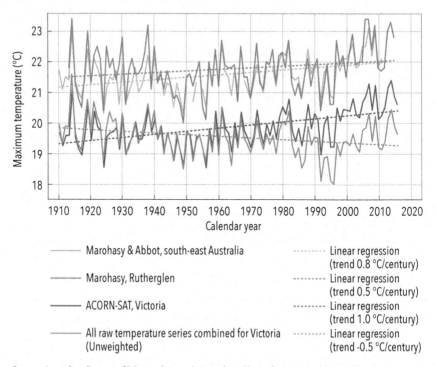

Source: Australian Bureau of Meteorology website at http://www.bom.gov.au/climate/data/ or http://www.bom.gov.au/climate/change/acorn-sat/#tabs=Data-and-networks.

Conclusion

Climate change is a topic of intense political interest. It is often reported that temperatures are increasing somewhat dramatically year on year, and that they are currently at unprecedented high levels. The official ACORN-SAT reconstruction for Victoria, as published by the Bureau, is consistent with this narrative and shows a dramatic recent rise in temperatures (see the red line in Figure 10.6), with an overall rate of warming of 1 °C per century (see column 1 in Table 10.1), and which is also consistent with widely-accepted global trends.

The ACORN-SAT reconstruction represents a combination of series of different lengths, from geographically different locations, that have been remodelling through the process of homogenisation and with complex area weightings applied. Because there are relatively large differences in mean maximum temperatures between weather stations (as shown in Figure 10.5), and large differences in mean maximum temperatures between drought and flood years (particularly at inland locations, as for example at Mildura, as shown in Figure 10.4), the particular combination of individual temperature series used at any one time, and the start date (note the difference between columns 2 and 3 in Table 10.1), can have a major effect on the rate of warming.

In other words, it is relatively easy to manipulate this key statistic; it is relatively easy to manipulate temperature trends.

Most Australian and international researchers rely exclusively on this ACORN-SAT record (for example, Coates et al. 2014), however, we recommend that more attention be paid to the raw data, which can be downloaded as both daily and monthly values from the Bureau website. This enables a richer and more realistic appraisal of the climate history of Victoria – and potentially of Australia more generally.

When the longest continuous series are combined in a very simple way, with a transparent system of area weightings (for example Marohasy

& Abbot 2016), the temperature trend for Victoria is only 0.3 °C per century (see column 2, in Table 10.1). This is probably close to the actual temperature trend for this part of Australia considering the period from 1887 to the present. However, Marohasy and Abbot (2016) suggest that in recent years the rate of warming has been closer to 2 °C per century.

It is a wrong assumption, inherent to mainstream climate science, that to assess climate variability and change it is necessary to work with only a subset of the data, and that the raw measurements must be first homogenised and area weighted.

In this study we have shown that the 289 individual maximum temperature series for Victoria show a very high degree of synchrony – they move in unison. This would *not* be the case if there were significant problems with the raw data. Furthermore, the presence of divergence, for example the steeper temperature increase for Melbourne, can be easily explained through the difference in local land use (i.e. the Urban Heat Island effect).

Simply combining all the temperature series to generate a mean annual value for Victoria will not though, give a realistic indication of climate variability and change. This is because the number and location of weather stations have changed, in particular the addition of series from cooler alpine regions since the 1980s.

Rather than applying an area weighting in an attempt to rectify this problem, it is our intention that future reconstructions comprise many longer temperature time series. This will be achieved by estimating historical temperature values through the application of artificial neural network technologies – including extending the alpine stations back in time. This would not only enable the calculation of more accurate temperature trends with less potential bias, but also provide rural and regional communities and industries with a better understanding of their climate history.

11 A Brief Review of the Sun–Climate Connection, with a New Insight Concerning Water Vapour

Dr Willie Soon & Dr Sallie Baliunas

Summary

This chapter gives a brief historical overview of the scientific study of the Sun–climate relationship. We then provide evidence for the existence of physical links between total solar irradiance (TSI) and Northern Hemisphere surface air temperature records from the Arctic, the US, Ireland and China, as well as the composite four-region record, as recently reported in Soon, Connolly and Connolly (2015). The results for TSI and conterminous US and Ireland temperature records are new. We use quality-controlled datasets for both TSI and surface temperature records avoiding issues that can arise from the contamination of temperature records by non-climatic factors. This is achieved by either adopting rural-only temperature datasets or urbanisation-corrected records. In this study, we show how TSI may be correlated with atmospheric water vapour. Left for future investigation is the development of a comprehensive theory of the climate and Sun–climate connection.

Early history

Early observations and speculations in classical Greece some 2300 to 2400 years ago, can be traced to Theophrastus (371–287 BCE) who suggested a connection between sunspots, and rain and wind (Soon and Yaskell 2003).

The Jesuit astronomer Giovanni Battista Riccioli (1598–1671) also wondered about a connection between sunspot activity and weather. Riccioli, in *Almagestum novum* published in 1651, noted the warm and dry weather in Italy during September of 1632 when sunspots were absent, in contrast to the cold June of 1642 when sunspots were prevalent. Galindo and Saladino (2008) recently pointed out a remarkable, early insight from the Mexican astronomer and meteorologist, Jose Antonio Alzate (1737–1799), who commented in 1784 that:

> The light conveyed by the Sun, efficiently influences vegetables and live-stock markets: this assertion should be present in the minds of those savants dealing with meteorological observations (none has implemented it). Pliny tells us in his 'Natural History' that, in times of Augustus, the Sun was dim and epidemics and food scarcity struck; modern physicists ascribe the mentioned dimness to the numerous sunspots that masked the solar disc at that time; from the year [17]69 I have been observing the Sun most days till the year [17]83, always numerous sunspots. In relation to the Eclipse of August 15, [17]84, I surveyed the Solar body using a telescope with much magnification and great clarity and confirmed that it was completely spotless. I continued observing daily up to 29 October, and no single spot appeared. Would the seasonal alterations experienced in Europe and here, depend partly on that cause? Only experience will tell.

With the benefits of modern measurements and equipment, the specific suggestion by Alzate of more dark sunspots leading to a dimmer Sun can now be rejected (see, for example, Scafetta and Willson 2014). For a more knowledgeable understanding of the Sun, fast-forward another 125 years, when Annie Maunder and Walter Maunder (1908), with their long experience of observing the Sun, remarked in *The Heavens and their Story* that:

> A ['great'] spot like that of February, 1892 is enormous by itself, but it is a small object compared to the sun; and spots of such size do not occur frequently, and last but a very short time. We have no right to expect,

therefore, that a time of many sunspots should mean any appreciable falling off in the light and heat we have from the sun. Indeed, since the surface around the spots is generally bright beyond ordinary, it may well be that a time of many spots means no falling off, but rather the reverse.

This acute observational fact alone tells us that a study of TSI variability, applied to a physical study of Sun–climate relations, must not be blindly equated to the study of sunspot count statistics *per se* (an important issue, which we will discuss again further, below).

In the English-speaking world, the idea and proposal concerning sunspots, weather and climate by William Herschel (1738–1822), in his twin papers published in *Philosophical Transactions of the Royal Society*, is popularly known and cited. These pioneering observations and remarks by Herschel (1801a, 1801b) have been considered both a positive support for and, further, deep scepticism of the physical reality of the Sun–weather–climate relationship. In his May 1801 paper, which is possibly less read and cited, he wrote:

> But we need not in the future be at a loss how to come at the truth of the current temperature of this climate, as the thermometrical observations, which are now regularly published in the Philosophical Transactions, can furnish us with a proper standard, with which the solar phenomena may be compared. ... [A]lthough I have, in my first Paper, sufficiently noticed the want of a proper criterion for ascertaining the temperature of the early periods where the sun has been recorded to have been without spots, and have also referred to future observations for shewing whether a due distribution of dry and wet weather, with other circumstances which are known to favour the vegetation of corn, do or do not require a certain regular emission of the solar beams ... For, if the thermometer, which will be our future criterion, should establish the symptom we have assigned, of a defective or copious emission of the solar rays, or even help us to fix on different ones, as more likely to point out the end we have in view, we may leave it entirely to others, to determine the use to which a fore-knowledge of the probable

temperature of an approaching summer, or winter, or perhaps or both may be applied; but still it may be hoped that some advantage may be derived, even in agricultural economy, from an improved knowledge of the nature of the sun, and of the causes, or symptoms, of its emitting light and heat more or less copiously.

We have quoted from both Alzate and Herschel in order to explain the limitations that existed to the study of the Sun–climate connection during their times. As well, it raises awareness about similar limitations in our own era; physical quantities are needed for a more quantitative and secure study, and to claim any relationship between the Sun, weather and climate. In the late eighteenth and early nineteenth centuries, thermometer records were neither widely available nor mature in their precision. As a result, Alzate and Herschel were forced to adopt the price of wheat or food as an indirect measure of meteorological and climatic fluctuations.

Thomas Jefferson (1743–1826), while concurring with Herschel's remark, was also calling for a firmer physical basis to study our climatic history. In marginal notes from his weather diary made at Monticello from 1 January 1810 to 31 December 1816, Jefferson says:

It is a common opinion that the climates of the several states of our union have undergone a sensible change since the dates of their first settlements; that the degrees of both cold and heat are moderated. The same opinion prevails as to Europe; if facts gleaned from history give reasons to believe that, since the times of Augustus Caesar, the climate of Italy, for example, has changed regularly at the rate of 1° of Fahrenheit's thermometer for every century. May we not hope that the methods invented in latter times for measuring with accuracy the degrees of heat and cold, and the observations which have been and will be made and preserved, will at length ascertain this curious fact in physical history?

Measurements remain a problem because, as we shall discuss and amplify later this chapter, even with the given thermometer records, the hardest tasks of any serious scientific study and analysis is to discern, and therefore remove, any contamination of the datasets by non-climatic factors. This problem is explained in detail by Connolly and Connolly (2014a; 2014b; 2014c). In addition, any Sun–climate study that relies strictly on sunspot-number records for quantitative assessment of solar activity level or strength may miss the correct physical and dynamic ranges of solar variability. It has long been warned, for example in Menzel (1959), that the 'zero' in sunspot numbers has an ambiguous meaning and value.

Before we consider these issues in more detail, let us briefly summarise findings from other contemporary studies of the Sun–climate connection.

Contemporary studies

In several previous publications, we have also shown and discussed the physical plausibility and relationship between TSI variations and various key climatic indicators and paleoproxies. Examples of these are the North Atlantic Meridional Overturning Circulation Index; Tropical Atlantic sea-surface temperatures; glacier mass and movements; zones and areal extent of tropical rain belts; and even waveguide teleconnection patterns, which have originated meteorologically, and where certain apparent correlations with zero-time lag as well as delayed responses of up to 50 to 100 years are all documented (Hormes et al. 2006; Soon 2009; Soon et al. 2011; Asmerom et al. 2013; Cruz-Rico et al. 2015; Yan et al. 2015). A more comprehensive discussion on the nomenclature of solar and climatic oscillations is given in Soon et al. (2014).

Recently, since the pioneering works discussed above, the scientific pursuit of the Sun–climate connection has been expanded, both in content and context. At the same time, it is openly admitted and discussed that recent studies are also not without controversies, with some having both technical strengths and weaknesses. Courtillot, Le Mouël and colleagues, in a series of papers (Le Mouël et al. 2009;

Courtillot et al. 2010; Kossobokov et al. 2010), introduced more advanced statistical techniques and various transformed climatic variables for addressing the non-stationary and non-linear nature of the solar forcing and climatic response relations. Van Loon et al. (2012) examined the multidecadal patterns of sea-level pressure and the associated quasi-stationary wave in the North Atlantic region, and suggested that this pattern could be related to the 50- to 100-year so-called Gleissberg–Yoshimura cycle (Solheim et al. 2012); while examining the relations between sunspot-cycle length and surface temperatures from Norway and the North Atlantic region offered the prediction of a significant temperature decrease during sunspot cycle 24 (that is, from about 2008 to 2018–2019).

Zhao and Feng (2014), from the Space Weather community, emphasised the solar activity and Earth–climate connection in terms of the multidecadal-centennial to bicentennial modulation. Many others are studying solar activity and weather and climate relations through charged-particle effects, including through incoming galactic cosmic rays and the global atmospheric electric circuit that are both in turn modulated by the solar wind (Svensmark & Friis-Christensen 1997; Soon et al. 2000; Yu 2002; Shaviv 2005; Harrison & Stephenson 2006; Usoskin et al. 2006; Miyahara et al. 2008; Harrison & Usoskin 2010; Kirkby et al. 2011; Tinsley 2012; Voiculescu & Usoskin 2012; Svensmark et al. 2013; Nicoll & Harrison 2014; Veretenenko & Ogurtsov 2104; Yu & Luo 2014; Zhou et al. 2014; Lam & Tinsley 2016; Prikyl et al. 2016).

The latest progress report in studying the cosmic-rays–cloud hypothesis by the CERN CLOUD experiment (Kirkby et al. 2016) has confirmed the importance of cosmic rays in elevating the natural background – in the absence of sulphuric acid that is known to be mostly an anthropogenic factor – of the biogenic emission of aerosol particles of 1.7 nanometers, or larger. However, there is still a long way to go to find a complete answer to the mystery of cloud formation in pristine and polluted environments.

Finally, the quasi-review paper by van Geel and Ziegler (2013) offered some perspective and context for solar activity and climatic covariations from the point of view of paleoclimate proxies. They concluded that the United Nations' (UN) Intergovernmental Panel on Climate Change (IPCC) reports have indeed underestimated the Sun's role in climate change. This conclusion challenges the community consensus-driven review on the Sun–climate connection recently published by Gray et al. (2010). In addition, Soon (2014) also independently reviewed the consensus Sun–climate discussion, as reported in the latest IPCC AR5 Working Group I report, and found serious misrepresentation, missing physical mechanisms, and political biases.

The difficulty of measuring TSI

The first issue that must be addressed in any solar TSI-climate study will concern the choice of measurement for TSI. All measurements are approximations of the total solar energy arriving at the Earth's atmospheric column and/or its surface, but there is no universally agreed measure. This issue is not widely understood, even among climate scientists.

It is shown in Soon, Connolly and Connolly (2015) that most of the often-quoted TSI records are based on the work of Wang, Lean and Sheeley (2005). These records have been subsequently used in the IPCC reports. They are very closely correlated to the sunspot-number records and, hence, are likely unable to capture our hypothesised, longer-term changes and modulation of solar irradiance rooted in various magnetic structures and complexes. This fact is consistent with the pioneering observations by Maunder and Maunder (1908) quoted earlier.

Indirect evidence supporting longer-term modulation of TSI, beyond the eleven-year cycle, such as sunspots variations, can be found by studying the changes in the fluid dynamics associated with the three-dimensional structures of magnetic dark spots, and related changes in solar convective zone properties (Hoyt 1979; Hathaway 2013; Bludova et al. 2014; Livingston & Watson 2015).

Figure 11.1 compiles seven proposed TSI reconstructions and the TSI index originally proposed by Hoyt and Schatten (1993), with updates (orange line top chart).

The diverse ranges of reconstructed TSI histories and values are explained mainly from the wide range of unknowns and uncertainties involved in the study of TSI beyond the direct observations by satellite-borne radiometers available only from 1979 to the present. The differences have been studied and discussed in detail in Soon, Connolly and Connolly (2015). In addition, as discussed above, TSI reconstructions that are based mainly on sunspot records alone may fail to fully capture the true amplitude and variation of the TSI index.

We chose the TSI reconstruction from Hoyt and Schatten (1993) because their work involves the most diverse types and ranges of proxy values for solar irradiance estimation – sunspot-cycle amplitude, sunspot-cycle length, solar equatorial-rotation rate, fraction of penumbral spots, and the decay rate of the approximate eleven-year sunspot cycle.

Their assumption is that each of those slightly different proxies will most likely capture some part of the underlying mechanism responsible for modulating the solar magneto-convection-induced processes that affect TSI. In an *á priori* sense, we note that magneto-fluid dynamical processes on the Sun need not strictly produce eleven-year-like cycles of high and near-zero sunspot numbers, as specified artificially by the paleoclimate modelling community (see Schmidt et al. 2011, 2012). In that regard, it is important to note that in the most recent work on

Figure 11.1 Different measures of total solar irradiance

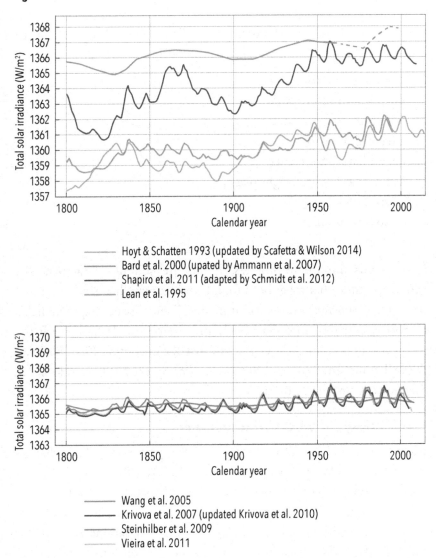

Hoyt & Schatten 1993 (updated by Scafetta & Wilson 2014)
Bard et al. 2000 (upated by Ammann et al. 2007)
Shapiro et al. 2011 (adapted by Schmidt et al. 2012)
Lean et al. 1995

Wang et al. 2005
Krivova et al. 2007 (updated Krivova et al. 2010)
Steinhilber et al. 2009
Vieira et al. 2011

The top chart shows the results from four studies suggesting relatively high solar variability since 1800. (The dashed blue line represents the work of Ammann et al. 2007, which was used to update Bard et al. 2000.) This contrasts with the bottom chart showing the results from low solar variability studies used as 'natural forcings' for the CMIP5 Global Climate hindcasts in the IPCC's Fifth Assessment Report (5AR).

Source: Adapted by permission from Elsevier – Soon, W, Connolly, R & Connolly, M, 'Re-evaluating the role of solar variability on Northern Hemisphere temperature trends since the 19th century', *Earth-Science Reviews*, vol. 150, pp. 409-452, copyright 2015.

physical modelling of TSI variability by Fontenla et al. (2011), up to nine solar magnetic features and structures were considered:

1. dark quiet-Sun inter-network
2. quiet-Sun inter-network
3. quiet-Sun network lane
4. enhanced network
5. plage (that is, not facula)
6. facula (that is, very bright plage)
7. sunspot umbra
8. sunspot penumbra
9. hot facula.

Despite this important progress, results are still unsatisfactory. The reason for this is that it has been found, and known, that even the 'quiet' part of the Sun may consist of 'unresolved' small-scale magnetic fields that vary, perhaps more than likely in a non-linear fashion, with the solar cycle in both mean strength and spatial distribution (Schuhle et al. 2000; Trujillo Bueno et al. 2004; Schnerr & Spruit 2011; Stenflo 2012; Stenflo & Kosovichev 2012).

Correlating TSI with surface temperatures

Many researchers have failed to find a relationship between TSI and changes in the Earth's surface temperatures because they used the wrong TSI proxy measure, and then chose a corrupted surface temperature record.

It is important when aiming towards a physical study of the Sun–climate connection to search for multidecadal covariations as a physical measure of the Sun's manifest activity. One relationship that has been explored is the TSI index in rural and urbanisation-corrected temperature datasets from the Northern Hemisphere. When high-quality surface temperature records are used with the Hoyt and Schatten TSI data, intriguing correlations can be identified (1993). Specifically, Soon

(2005) and Soon (2009) found a strong apparent correlation for the Arctic; Soon et al. (2011) found a similar correlation with China; Soon and Legates (2013) also found a correlation between TSI and trends in the 'Equator-to-Pole (Arctic) Temperature Gradient'.

How can one be sure, however, that those temperature records have not been contaminated with non-climatic effects and factors? If they have been contaminated, how does one quantify the non-climatic components, and therefore, remove those factors from the raw or unmodified datasets?

Figure 11.2 shows the temperature anomalies for key regions after these questions have been answered in detail, with the workings described in Soon, Connolly and Connolly (2015). The statistical correlations shown in Figure 11.2, considering how diverse the geographical areas and climatic zones that are covered, are very impressive.

Correlating TSI with water vapour concentrations

A so far much neglected key variable that is important in understanding ocean–land–atmosphere climate systems is the total atmospheric water vapour. However, a direct measurement of this metric is not available for a period of 100 years, and so the record must be constructed by using indirect deduction from the so-called climate reanalyses study (Hersbach et al. 2015; Kobayashi et al. 2015; Poli et al. 2016).

Figure 11.3 shows the correlation between the TSI index of Hoyt and Schatten and the total column water vapour content, as deduced from the European Centre for Medium-Range Weather Forecasts' model outputs. The relationship is close and meaningful, especially for the multidecadal variation and modulation. However, we note that the inter-annual changes are more closely related to internal oscillating components of the climate system, such as the El Niño–Southern Oscillation (ENSO) or the North Atlantic Oscillation (NAO) factors. The close correlation, perhaps, reflects a real physical relationship between

Figure 11.2 Comparing TSI and Northern Hemisphere temperatures

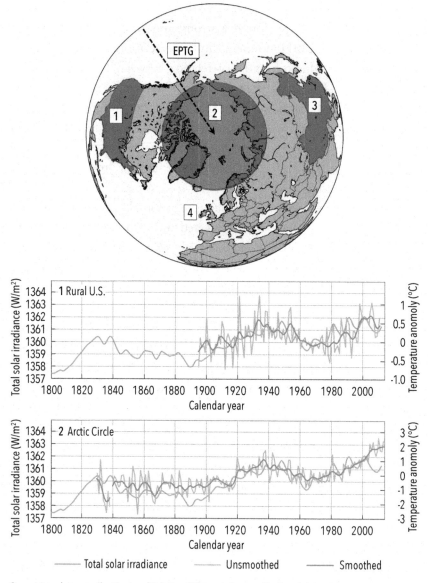

Comparison between the Hoyt and Schatten TSI trends (orange line) and the various temperature indices (blue lines) from the Northern Hemisphere regions (1. Rural US, 2. Arctic Circle, 3. China, 4. Rural Ireland, and the Northern Hemisphere composite record), as well as the Northern Hemispheric Equator-to-Pole temperature gradients (EPTG).

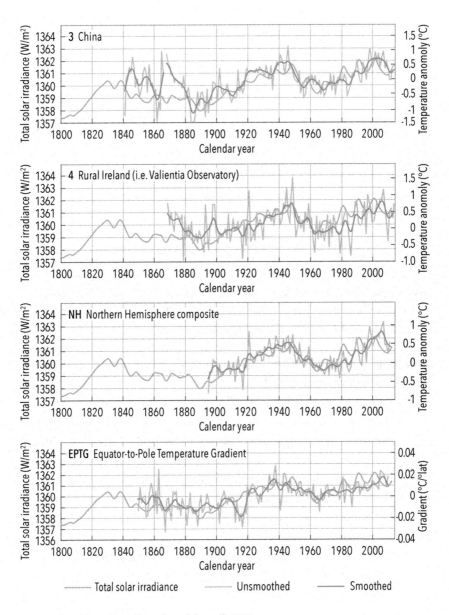

Source: Adapted from Soon, Connolly and Connolly 2015.

Figure 11.3 Comparing TSI and water vapour trends

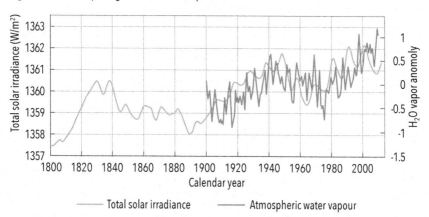

— Total solar irradiance   — Atmospheric water vapour

Comparison between the Hoyt and Schatten TSI trends (orange line) and the total atmospheric water vapour content anomaly (over ice-free region of the Earth's oceans) based on the twentieth-century climate reanalyses outputs from the climate model of the European Centre for Medium-Range Forecasts.

Source: Data courtesy of Drs A Simmons and H Hersbach.

sea-surface temperatures and the atmospheric water vapour, as convincingly emphasised in Hersbach et al. (2015).

If the results from Figure 11.2 and Figure 11.3, both in terms of physical correlations and mechanisms, can be fully confirmed, we will then have very clear empirical evidence for supporting a Sun–climate connection on multidecadal timescales. This is in the sense that it is the TSI-induced changes in modulating the direct thermal impact on the surface heat budget, as well as the amplification factor from the covarying atmospheric water vapour that best explained all the available observed climatic records.

In contrast, this fact will imply that the role of rising atmospheric CO_2 may be, at most, minimal when explaining the surface temperature and the atmospheric water vapour records. Indeed, in Soon, Connolly and Connolly (2015), we went a step further – by using the time series in combination with the atmospheric CO_2 forcing – and found that the quantitative constraints for the Earth's 'climate sensitivity' to the doubling of atmospheric CO_2 was anywhere between 0.4 to 2.5 °C. The high value is obtained by assuming, without any strong supporting arguments, that the Northern Hemisphere derived composite temperature is driven foremost by rising atmospheric CO_2 content alone. The lower value of 0.4 °C for CO_2-doubling climate sensitivity is obtained by assuming the observed empirical correlation shown in Figure 11.2 is valid, and then fitting only a CO_2 forcing curve to the residual temperature trend after accounting for the TSI factor.

Conclusion

It is hoped that the study of Sun–climate relations will remain true to science and the scientific method, and be able to be better interpreted using high-quality data.

We are, perhaps, moving slowly towards a unifying theory of climate (Lions et al. 1993; Monin & Shishkov 2000; Li & Wang 2008; Essex 2011, 2013; Palmer et al. 2014). This will be further strengthened when there is a better appreciation of the socio-political factors, which currently obscure a better understanding of the Sun–climate connection. This would be helped by a proper definition of 'climatology' (for example, Landsberg 1972; Gutmann 1989; Arguez & Vose 2011).

In Soon, Connolly and Connolly (2015) the importance of using uncontaminated surface temperature datasets is explained, along with reasons for choosing the Hoyt and Schatten TSI-record. The question about the veracity of TSI being adopted for such a study of the Sun–climate connection must necessarily remain open and be answered in time by solar physics. Even more importantly is the fact that several other promising mechanisms, especially those involving the charged particles and electric-circuit phenomena outlined above, must all be fully examined and integrated into a more comprehensive framework, which should, ultimately, give us the correct superposition rules for the Sun–Earth climatic connection in multiscale spatial and temporal domains.

12 The Advantages of Satellite-Based Regional and Global Temperature Monitoring

Dr Roy W Spencer

The monitoring of regional and global temperature trends is controversial, partly because our temperature measurement systems were designed to observe very large day-to-day temperature changes, not the very small rates of approximately 0.25 °C per decade of global warming expected by many climate researchers. Since measured warming trends have generally been less than this expectation, it becomes even more difficult to say with any accuracy just how fast the climate system has warmed.

There are three basic methods now used to monitor regional and global temperature changes:

- surface-based thermistors
- radiosondes (weather balloons)
- satellites.

Each has its advantages and disadvantages, and, hopefully, we know more about the climate system using all three systems together rather than just one.

Surface-temperature records, which extend back to the mid-1800s, were originally recorded from liquid-in-glass thermometers. Over recent decades these thermometers, which were read visually, have been mostly replaced with electronic thermistors. Their coverage of the Earth is still somewhat patchy, being mostly restricted to inhabited land areas.

Most of the thermometer data shows spurious warming effects because the thermometers are often proximal to man-made structures. This 'urban heat island' (UHI) effect is difficult to quantify and correct for.

Radiosondes have been operating since about the 1950s. Their coverage is even sparser than surface thermometers, but they have the advantage of measuring the atmosphere at many different altitudes, which is useful to better understand climate variations. These instruments are primarily used to provide input into computerised weather prediction models, but are now also used for climate monitoring, despite changes in instrument design over the years.

The monitoring of global temperatures with satellites has been possible since 1979, using microwave radiometers that measure the thermal emission of microwave energy by oxygen in the atmosphere (Spencer & Christy 1990).

The intensity of the measured microwave energy is directly proportional to the temperature of the air. Despite the relatively short data record, satellites have the advantage of near-global daily coverage, even in remote areas of the world where there are no surface thermometers. The satellites measure the average temperature of deep layers of the atmosphere, and are most directly comparable to the radiosonde measurements, which can be vertically averaged to approximate the same layer(s) as the satellites measure.

While only a dozen or so satellites have been performing this monitoring in the last four decades, each one measures the average temperature of 25,000 km^3 of atmosphere every 1/20 of a second; as it travels across the Earth it then samples nearly every cubic kilometre of atmosphere every day. An obvious advantage is that satellite measurement of global temperatures provides near-complete coverage of the world. While small areas around the poles are difficult to measure, the satellites still cover 98% of the Earth.

While the satellite measurement of microwave radiation sounds like a rather indirect way to measure air temperature, it should be noted that since electronic thermistors replaced liquid-in-glass thermometers, most surface-based thermometers today, and all radiosondes, actually measure electrical resistance, which is also indirect. Liquid-in-glass thermometers (mercury or alcohol) are seldom used anymore, since they require visually estimating the temperature by measuring the expansion of the liquid in a glass tube. Using radiation to measure temperature is also commonly done with infrared devices, including with the instruments used by doctors to measure your body temperature.

The satellite measurements of the atmosphere are calibrated by measuring the cosmic background radiation from deep space – which is near absolute zero in temperature – and a warm calibration target within the instrument (Robel & Graumann eds. 2014) – the temperature of which is monitored with redundant platinum resistance thermometers that are laboratory-calibrated and highly accurate.

Two independent research groups provide the most widely referenced global temperature datasets: the University of Alabama in Huntsville (UAH) (Earth System Science Center 2017); and Remote Sensing Systems (RSS) (Mears & Wentz 2009), a private research firm in Santa Rosa, California. The products produced by the two groups give very similar results, despite some differences in data processing, data adjustments, and the mixture of satellites used.

The nature of the satellite measurements is different from, but complementary to, the surface thermometer data. Unlike the surface data, the satellites measure an average temperature over a fairly deep layer of the atmosphere, from the surface up to about 10 km in altitude. This layer is strongly coupled to the surface on regional and larger spatial scales, and on monthly and longer time scales. That means warming (or cooling) at the surface should be reflected in similar responses in deep-layer temperatures.

In fact, deep-layer temperature is predicted to warm faster than surface temperature in a warming world (Solomon et al. ed 2007), and therein lays the controversy over satellite temperatures: satellites actually show slower warming than is occurring at the surface, rather than faster.

Global-average warming trends

Until the 2015–16 El Niño warm event arrived, the satellite temperature trend averaged over the Earth had been near zero for approximately eighteen years, since about 1997. The monthly global average temperature variations of the lower troposphere since 1979 for the UAH dataset are shown in Figure 12.1.

During 2016, temperatures rapidly cooled again as a weak La Niña became established in the Pacific Ocean. While one can quibble over whether global warming has actually 'stopped' since 1979, there is no

Figure 12.1 Monthly global average temperature anomalies

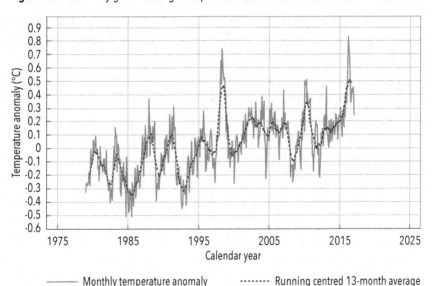

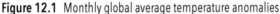

Source: UAH Satellite-based temperature of the global lower atmosphere, http://vortex.nsstc.uah.edu/data/msu/v6.0/tlt/uahncdc_lt_6.0.txt.

question that the slowdown in warming was unexpected, and (as we shall see) unexplainable by current computerised climate models that are relied on for global warming forecasts and energy policy.

Despite this recent 'pause' or 'hiatus' in warming, it can be seen in Figure 12.1 that there has still been a weak warming trend since 1979, about +0.14 °C per decade. The question is, just how significant is this? If it continued, it would be about 1.4 °C in 100 years, which would probably be easily adapted to by humans and the natural world.

But this slow rate of warming is not what is expected by many climate researchers. Significantly, the satellite-indicated warming rate is only about 50% of the warming rate predicted by the computerised climate-model projections tracked by the United Nations' (UN) Intergovernmental Panel on Climate Change (IPCC). If we compare the warming predicted by the climate models to the satellite-measured warming, we see that they have diverged since about 1998 (see Figure 12.2). Note that the radiosonde measurements, when averaged to match the same atmospheric layer measured by the satellites, also suggest a disagreement with the climate models.

The disagreement is even greater in the tropics (not shown). While an average of 102 climate models is shown in Figure 12.2 to indicate the average behaviour of climate models in general, virtually none of the individual models projected such a slow rate of warming as has been observed by satellites and radiosondes. Generally, 95% to 97% of the models overestimate the warming trend.

Since the climate models provide the basis for changes in energy policy to reduce the emission of greenhouse gases – the supposed cause of warming – this makes the satellite measurements critical to the discussion of just how necessary such energy policy changes really are.

What are the possible reasons for the observations and climate models not agreeing more? First, the satellite observations could be in error. But that would also require the radiosonde (weather-balloon) data – most

Figure 12.2 Climate models versus satellite and radiosonde observations

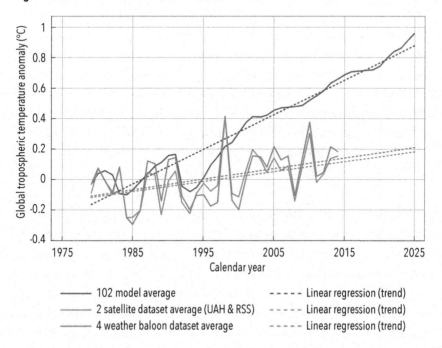

Source: adapted from Christy, JR, 2016, 'Testimony to the U.S. House Committee on Science, Space & Technology', 2 February, viewed 28 March, 2017, http://docs.house.gov/meetings/SY/SY00/20160202/104399/HHRG-114-SY00-Wstate-ChristyJ-20160202.pdf

analyses of which are supportive of the satellite measurements as well – to have the same error. So, barring the unlikely coincidence that both the satellite and radiosonde analyses are in error by the same amount, we will assume they are largely correct.

This leaves the model forecasts as the source of the error. One possibility is that there has been a temporary, but unknown, cooling influence, which will eventually end with rapid warming resuming at some point in the future. In other words, we just haven't waited long enough for the full strength of global warming to be realised and the models will eventually be proved correct. For example, an increase in heat storage by the deep ocean, which is very cold, might explain the discrepancy. However, the rate of deep-ocean warming has also been less than the models predict, especially when one factors in the effects of El Niño and La Niña (Spencer & Braswell 2014).

The more likely problem with the models is that they are tuned to be too sensitive to our greenhouse-gas emissions. There are many uncertainties about how climate system elements, such as clouds, will change with global warming, either acting to reduce warming or enhance warming, and these uncertainties in our understanding necessarily find their way into the climate models. The models are tuned so that changes in clouds, water vapour, and other elements amplify the relatively weak direct warming from increasing carbon dioxide (CO_2). If the climate system actually acts to dampen such warming, then the climate-model projections of warming could easily be a factor of two times too large.

While it is true that most of the physics contained in climate models is well understood, these so-called 'feedbacks', especially those due to clouds, involve large uncertainties and are included in the models in rather crude and approximate ways. Since it is the feedback effects that largely determine just how much warming and associated climate change a model will produce, the uncertainties about their values directly translate into uncertainties in projections of future warming.

Regional warming trends

When we examine how different parts of the world have warmed since 1979, we see that land areas have warmed somewhat faster than the ocean areas (see Figure 12.3). This is what we would expect, since warming of the oceans (no matter the cause) is mixed downward much more rapidly and so is spread over much more mass than it is over land. For example, the same amount of heat from a stove will cause a much smaller rise in temperature for 25 cm of water in a pot than it would for 2 cm of soil.

Figure 12.3 Regional linear temperature trends

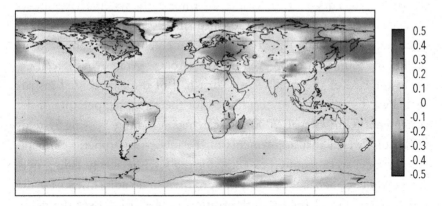

Source: Calculated from the UAH Satellite-based temperature of the global lower atmosphere, January 1979 through to January 2016.

Also, the high northern latitudes have warmed the most since 1979, and as one progresses southwards, the rate of warming has been less, reaching near-zero temperature change around Antarctica.

Regional differences in temperature trends are to be expected, even over the 37-year period represented in Figure 12.3, just due to variability associated with El Niño, La Niña, and other natural modes of climate variability. This is why 'global warming' cannot be expected to be clearly seen unless averages are computed over large regions, even over hemispheres or the whole Earth. Again, this is because the expected global warming signal is so small compared to the natural variations in the climate system.

How accurate are the satellite measurements versus thermometers?

There are no simple answers regarding the accuracy of temperature measurements when it comes to monitoring trends, because we do not have a good way of determining all of the sources of errors over several decades. Thermometers, radiosondes, and satellites were designed to measure temperature changes of many degrees on a daily basis, not hundredths of a degree over decades. All three of the systems have undergone changes over time, which leads to small spurious effects that can affect long-term trend calculation. All measurements require adjustments for known sources of error, even if those errors aren't known perfectly.

In the case of the satellites, the first half of the record used instruments we discovered must be adjusted for the actual temperature of the instrument (Christy et al. 2000). Researchers at Remote Sensing Systems (RSS) discovered that adjustments of the lower-tropospheric temperature product were required to allow for the satellite slowly falling over time (Wentz & Schabel 1998). Both groups also do adjustments for the satellites because of the local observation time drifting over the years, called a diurnal adjustment.

What we do know is that the adjustments made to the satellite data are generally small (in the order of 0.1 °C or less), while adjustments to the thermometer and thermistor data are typically much larger (often 1 °C or more). So, it is possible that the satellite data are more accurate for assessing regional and global temperature trends, although no one has been able to prove this. It should be remembered that, while the satellite data are adjusted for all known effects, no one has yet found a way to remove the very local UHI warming effect from the thermometer and thermistor data. For the most part, such spurious warming effects are indistinguishable from human-caused global warming, which is a problem when the measurements are being used to assess regional and global climate variability and change caused by increasing concentrations of greenhouse gases.

One way to demonstrate the existence of the UHI effect in thermometer data is to compare nearby pairs of thermometers which are in different population environments. To do this, one does not need to classify a thermometer as either 'urban' or 'rural', we only need to know the population density of their immediate surroundings. For example, in Figure 12.4 I used the average temperature differences between thermometer pairs located in different population density environments less than 150 km apart, and averaged all of the results from all temperature reporting stations in the world during the year 2000, computing the average local warming as population increases.

As can be seen from Figure 12.4, and as found by previous investigators, the strong warming effects exist even at low population densities. In other words, even minimum amounts of man-made structures have a fairly large impact on temperatures, and only truly rural sites are immune from UHI effects. These are effects that do not exist in the satellite data because they are restricted to a very shallow layer of air near the surface, and only in urban areas.

Figure 12.4 The global average UHI effect

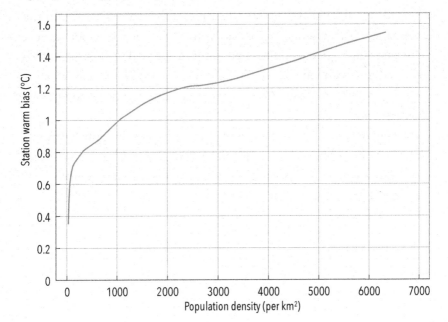

Source: Calculated for the year 2000, based on 16,651 unique station pairs less than 150 km apart.

While the UHI effects have been claimed to have been handled by statistical adjustment schemes that match rural to urban data, it has never been demonstrated that the contaminated urban data have been corrected to match the rural data, rather than the other way around.

Conclusion

The monitoring of global temperatures by satellites has certain advantages over monitoring with thermometers, most notably: near-global coverage; immunity from UHI effects; and smaller corrections needed to the data.

While the satellites do not measure near-surface temperatures 'where people live', the layer of the lower atmosphere that they do measure is where we expect the strongest global warming signal. For this layer, the measured rates of warming since 1979 have been about 50% of that expected by climate models for the same layer, which is a significant discrepancy. The discrepancy is likely the fault of the models, since radiosonde data agree with the satellites. It might well be that the models are tuned to be too sensitive to CO_2 emissions, and that global warming resulting from increasing CO_2 will not be as serious a problem as many scientists believe.

It remains to be seen in the coming years whether the global warming 'pause' will re-establish itself after the recent 2015–16 El Niño warm event is followed by the 2016–17 La Niña cool event, or whether a weak warming trend will continue.

13 Carbon Dioxide and Plant Growth
Dr Craig D Idso

Introduction

Scientists have been concerned about the biospheric consequences of Earth's rising atmospheric carbon dioxide (CO_2) concentration for well over a century. Primarily driven by gaseous emissions from the burning of fossil fuels, such as coal, gas and oil, the air's CO_2 content has risen from a mean concentration of about 280 parts per million (ppm) in 1800 to a value of approximately 405 ppm today. If current fuel consumption trends continue, it is projected that the planet's atmospheric CO_2 concentration may reach upwards of 800 ppm by the end of this century (Stocker et al. eds. 2013). It is thus only natural to wonder how Earth's climate and its plant and animal life will respond to this global environmental change. Although much of the scientific and public debate focuses on the climatic aspect of the planet's rising CO_2 concentration, this chapter discusses several biological implications of this very real and ongoing phenomenon for Earth's plants.

Basic plant responses to atmospheric CO_2 enrichment
Increased productivity

Perhaps the best-known consequence of enriching the air with CO_2 is that plant productivity is stimulated. This growth enhancement occurs because, at a fundamental level, CO_2 is the basis of almost all life on Earth. It is the primary raw material used by plants during the process of photosynthesis to produce the organic matter out of which they construct their tissues. Consequently, literally thousands of laboratory and field experiments conducted over the past several decades have demonstrated that the more CO_2 there is in the air, the better plants will grow (Idso et al. 2014).

For herbaceous plants, a 300-ppm increase in the air's CO_2 content typically raises their productivity by about one-third; and this positive response occurs in plants that use all three of the major biochemical pathways (C_3, C_4, CAM) of photosynthesis. Therefore, with more CO_2 in the air, the productivity of nearly all crops rises (see Table 13.1), as they produce more branches and tillers, more and thicker leaves, more extensive root systems, and more flowers and fruit (Idso & Idso 2011). It should come as no surprise, therefore, that the father of modern research in this area, Sylvan H Wittwer (1997), has stated that 'it should be considered good fortune that we are living in a world of gradually increasing levels of atmospheric CO_2', and that 'the rising level of atmospheric CO_2 is a universally free premium, gaining in magnitude with time, on which we can all reckon for the future'.

The growth response of woody plants to atmospheric CO_2 enrichment has also been extensively studied, and their response to increasing CO_2 is generally larger than that observed for herbaceous plants. Reviews of numerous individual woody plant experiments, for example, have revealed a mean growth enhancement of about 50% for an approximate doubling of the air's CO_2 content, which amounts to around one and a half times the response of non-woody herbaceous plants (Poorter 1993; Ceulemans & Mousseau 1994; Wullschleger et al. 1995, 1997).

Although much less studied than terrestrial plants, many aquatic plants are also known to be responsive to atmospheric CO_2 enrichment, including unicellular phytoplankton and bottom-rooted macrophytes, for both freshwater and saltwater species (Idso et al. 2014). Therefore,

Table 13.1 Mean percentage biomass increase

Crop	% Biomass change	Crop	% Biomass change
Sugar cane	34.0	Rye	38.0
Wheat	34.9	Plantains	44.8
Maize	24.1	Yams	47.0
Rice, paddy	36.1	Groundnuts, with shell	47.0
Potatoes	31.3	Rapeseed (Canola)	46.9
Sugar beet	65.7	Cucumbers and gherkins	44.8
Cassava	13.8	Mangoes, mangosteens and guavas	36.0
Barley	35.4	Sunflower seed	36.5
Vegetables fresh	41.1	Eggplants (aubergines)	41.0
Sweet potatoes	33.7	Beans, dry	61.7
Soybeans	45.5	Fruit fresh	72.3
Tomatoes	35.9	Carrots and turnips	77.8
Grapes	68.2	Other melons (inc. cantaloupes)	4.7
Sorghum	19.9	Chillies and peppers, green	41.1
Bananas	44.8	Tangerines, mandarins and clementines	29.5
Watermelons	41.5	Lettuce and chicory	18.5
Oranges	54.9	Pumpkins, squash and gourds	41.5
Cabbages and other brassicas	39.3	Pears	44.8
Apples	44.8	Olives	35.2
Coconuts	44.8	Pineapples	5.0
Oats	34.8	Fruit, tropical, fresh	72.3
Onions, dry	20.0	Peas, dry	29.2
Millet	44.3		

Source: Idso, CD 2013.

there is probably no category of photosynthesising plant that does not respond in a positive manner to atmospheric CO_2 enrichment, or is not likely to benefit from the ongoing rise in the air's CO_2 content.

Enhanced water-use efficiency

Another major consequence of atmospheric CO_2 enrichment is that plants exposed to elevated levels of atmospheric CO_2 generally do not open their leaf stomatal pores – through which they take in CO_2 and give off water vapour – as wide as they do at lower CO_2 concentrations. In addition, at higher levels of atmospheric CO_2, they tend to produce less of these pores per unit area of leaf surface. Both these changes tend to reduce most plants' rates of water loss by transpiration; and the amount of carbon they gain per unit of water lost – or water-use efficiency – therefore, typically rises, greatly increasing their ability to withstand drought. And with fewer and smaller stomatal openings, plants exposed to elevated levels of atmospheric CO_2 are also less susceptible to damage by noxious air pollutants – including ozone and oxides of nitrogen and sulphur – that gain entry into plants via these portals.

In addition to the obvious benefits that CO_2-induced increases in plant water-use efficiency mean for both irrigated and dry-land agriculture, there are equally important consequences for Earth's natural ecosystems. As the air's CO_2 content continues to rise, for example, plants will be able to grow and reproduce where it has previously been too dry for them to exist. Consequently, terrestrial vegetation should become more robust and begin to win back lands previously lost to desertification. Simultaneously, the greater vegetative cover of the land produced by this phenomenon should reduce the adverse effects of soil erosion caused by the ravages of wind and rain. Evidence that such is indeed occurring was presented several years ago by Trimble and Crosson (2000) who reported that 'available field evidence suggests declines of soil erosion, some very precipitous, during the past six decades'.

Greater plant productivity should also lead to increases in soil organic matter, which will likely produce even further benefits for the planet's plant life. In the words of Wallace (1994), soil organic matter:

> stops soil erosion, it supplies nutrients, it is a buffer against pH change, it holds water, it increases the cation exchange capacity ... [which protects against leaching and loss of nutrients], it decreases compaction, it stores nutrients from season to season, it makes soil warmer in the spring, it makes soils easier to till especially when slightly too wet, it makes inputs more valuable, it protects against plant diseases, it gives better-aerated more permeable soil, it protects against heavy metal and salt toxicities, it detoxifies pesticides and prevents their leaching, it is a storage mechanism for excess atmospheric CO_2, it gives high yields, it promotes microbial breakdown of toxic substances, it makes it possible to grow acid loving plants, it supports microorganisms that recycle nutrients, and it promotes soil formation.

Creatures that live in the soil, such as earthworms, are also greatly stimulated by additions of organic matter (Rogers et al. 1994); and an increase in their activity would likely lead to the creation of much new soil, while at the same time improving the fertility, structure, aeration and drainage of existing soils (Edwards 1997). These improvements, in turn, would likely boost plant productivity higher yet, putting still more organic matter into the soil, and so on (Idso 1991), with the several phenomena reinforcing and further enhancing each other in a way that lifts the biosphere to a new level of activity (Idso 1992).

Amelioration of environmental stresses

Atmospheric CO_2 enrichment has also been shown to help ameliorate the detrimental effects of various environmental stresses on plant growth and development (Idso et al. 2014). A high level of atmospheric CO_2 has been shown to help reduce the deleterious effects of high soil salinity, high air temperature, low light intensity and low levels of soil fertility. Additionally, it has been demonstrated to reduce the severity of low temperature stress, oxidative stress, and the stress of herbivory. In fact, the percentage growth enhancement produced by an increase in the air's CO_2 concentration is generally *greater* under stressful and resource-limited conditions than it is when growing conditions are ideal.

Perhaps the most significant of these stress-reduction effects is that pertaining to high air temperature, because it is global warming that tops the list of environmental concerns in most countries of the world. In this regard, there is a commonly held belief that temperatures could rise so high and so rapidly that many plants and trees, in particular, may not be able to migrate fast enough to keep up with the poleward-shifting thermal environments to which they are currently accustomed. However, the beneficial effects of atmospheric CO_2 enrichment typically *rise* with an increase in air temperature. In fact, for a 300-ppm increase in the air's CO_2 content, the mean CO_2-induced growth enhancement from 42 experiments analysed by Idso and Idso (1994) rose from a value of zero at 10 °C to a value of 100% at 38 °C.

This increase in CO_2-induced plant growth response with increasing air temperature arises from the negative influence of high CO_2 levels on the growth-retarding process of photorespiration, which can 'cannibalise' 40 to 50% of the recently produced photosynthetic products of C_3 plants. Since this phenomenon is more pronounced at high temperatures, and as it is ever-more inhibited by increasingly higher atmospheric CO_2 concentration, there is an increasingly greater potential for atmospheric CO_2 enrichment to benefit plants as air temperatures rise.

A major consequence of this phenomenon is that the optimum temperature for plant growth generally rises when the air is enriched with CO_2. For a 300-ppm increase in the air's CO_2 content, in fact, biochemical theory suggests that the temperatures at which most C_3 plants grow best will rise by about 5 °C (Long 1991); and several experimental studies have shown that the optimum temperature for growth in such plants typically rises by that amount or more for a 300-ppm increase in atmospheric CO_2 concentration (Bjorkman et al. 1978; Nilsen et al. 1983; Jurik et al. 1984; Seemann et al. 1984; Harley et al. 1986; Stuhlfauth & Fock 1990; McMurtrie et al. 1992).

These observations are of great significance, because an increase of this magnitude in optimum plant growth temperature is even larger than the largest air temperature rise predicted to result from a 300-ppm increase in atmospheric CO_2 concentration (Stocker et al. eds. 2013). Hence, even the most extreme global warming envisioned by the IPCC would probably not adversely affect the majority of Earth's plants; this is because fully 95% of all plant species are of the C_3 variety. In addition, the C_4 and CAM plants that make up the rest of the planet's vegetation are already adapted to Earth's warmer environments, which are expected to warm much less than the other portions of the globe; yet, even some of these plants experience elevated optimum growth temperatures in the face of atmospheric CO_2 enrichment. Consequently, it is doubtful that a CO_2-induced global warming would produce a massive poleward migration of plants seeking cooler weather; for the temperatures at which nearly all plants perform at their optimum would likely rise at the same rate (or faster than) and to the same degree as (or higher than) the temperatures of their respective environments (Idso 1995). Hence, there would be absolutely no need for plants to seek out cooler climates in a warming world where atmospheric CO_2 concentrations were also rising.

At even higher temperatures that are normally lethal to plants, atmospheric CO_2 enrichment has also been proven to be of great worth, because sometimes it can mean the difference between plants living or dying. A high CO_2 level, for example, has been demonstrated to enable plants to maintain positive leaf carbon exchange rates when similar plants growing under the ambient CO_2 concentration exhibited negative rates that led to their demise. Likewise, an elevated atmospheric CO_2 level tends to protect plants against the severe desiccation that often accompanies high temperatures.

Biospheric impacts of atmospheric CO_2 enrichment

The preceding material clearly indicates the potential for Earth's biosphere to reap numerous benefits as a result of the ongoing rise in the air's CO_2 content, many of which have already been realised to one degree or another. A few of these responses are highlighted in the sections below.

Agricultural ecosystems

One of the clear implications of the demonstrated ability of atmospheric CO_2 enrichment to enhance plant growth and development, even in the face of limited resources and the presence of debilitating environmental stresses, is that agricultural productivity should already be on the rise throughout the entire world, as a result of the aerial fertilisation effect produced by society's continually increasing emissions of CO_2.

This consequence has long been recognised by agricultural scientists. Nearly three decades ago, Gifford (1979) suggested that a significant part of the upward trend of Australia's wheat yields during the period 1958–1977 'may be due to atmospheric [CO_2] change rather than to managerial or genetic causes'. Three years later, Wittwer (1982) noted that 'the 'green revolution' has coincided with the period of recorded rapid increase in concentration of atmospheric CO_2, and it seems likely that some credit for the improved yields should be laid at the door of the CO_2 build-up.' Five years after that, Allen et al. (1987) gave some actual numbers for soybeans, concluding that their yields 'may have increased by 13%' from about 1800 AD to the time of their analysis due to a global CO_2 increase.

Similar sentiments have been expressed by other scientists in the years that have followed, with many of them looking forward to the future with great anticipation of the many CO_2-induced benefits that are likely to accrue to agriculture. In this regard, one recent study provided a quantitative estimate of the direct monetary impact of rising atmospheric CO_2 enrichment on both historic and future global crop production

(Idso 2013). The results of that analysis indicated that the annual total monetary value of this benefit grew from US$18.5 billion in 1961 to more than US$140 billion by 2011, amounting to a total sum of US$3.2 trillion during the 50-year period from 1961 to 2011. In projecting the monetary value of this positive externality, it was further determined that rising atmospheric CO_2 will likely bestow an additional US$9.8 trillion on global food production between 2012 and 2050. Perhaps more important, however, is the realisation that without the estimated future CO_2-induced productivity enhancements on agriculture, world food supplies would likely fall short of world food demand by the year 2050 (Idso & Idso 2000).

A future of global food insecurity, however, does not have to occur. Consider, for example, the fact that rice is the third most important global food crop, accounting for just more than 9% of all global food production. The average growth response of rice to a 300-ppm increase in the air's CO_2 concentration is around 37.2% (Center for the Study of Carbon Dioxide and Global Change n.d.). However, data obtained from De Costa et al. (2007), who studied the growth responses of sixteen different rice genotypes, has revealed CO_2-induced productivity increases ranging all the way from –7% to a whopping +263%. Therefore, if countries learned to identify which genotypes provided the largest yield increases per unit of CO_2 rise, and then grew those genotypes, it is quite possible that the world could collectively produce enough food to supply the needs of all its inhabitants, staving off the crippling food shortages that are projected to result in just a few short decades from now in consequence of the planet's ever-increasing population.

In summing up the potential benefits of rising atmospheric CO_2 for agricultural ecosystems, Wittwer (1995) stated some two decades ago:

> The rising level of atmospheric CO_2 could be the one global natural resource that is progressively increasing food production and total biological output, in a world of otherwise diminishing natural resources of land, water, energy,

minerals, and fertilizer. It is a means of inadvertently increasing the productivity of farming systems and other photosynthetically active ecosystems. The effects know no boundaries and both developing and developed countries are, and will be, sharing equally.

Forest ecosystems

Convincing evidence also exists for the effectiveness of the aerial fertilisation effect of atmospheric CO_2 enrichment in enhancing the productivity of Earth's forested ecosystems (Loehle et al. 2016). Many studies of historical trends of forest productivity have shown tree growth the world over to have risen together with the upward trend in the air's CO_2 content that began with the dawning of the Industrial Revolution, with some of the studied tree species actually doubling their growth rates over this time period (Idso 1995). This phenomenon is evident even in highly productive *tropical* forests; and it has been demonstrated to have accelerated in recent decades (Idso et al. 2014).

Concomitantly, multiple studies from every corner of the globe have revealed that the *ranges* of forests have also increased, particularly where they have not been curtailed by the actions of man (Idso et al. 2014). Both of these CO_2-induced phenomena are manifestations of the unprecedented botanical transformation of the planet that is lifting Earth's biosphere out of the low-productivity state it has occupied for the past six to eight million years of uncommonly low atmospheric CO_2 concentrations and restoring it to the high-productivity state that it occupied over the prior 400-million-year period of much higher atmospheric CO_2 concentrations.

The entire biosphere

In considering the plant productivity gains that result from the *aerial fertilisation effect* of the ongoing rise in atmospheric CO_2, plus its *transpiration-reducing effect* that boosts plant water-use efficiency, along with its *stress-alleviating effect* that lessens the negative growth impacts of resource limitations and environmental constraints (such as high temperatures), the world's vegetation possesses an ideal mix of abilities to reap tremendous benefits in the years and decades to come. And based on a multitude of real-world observations, that future is now. Evidence from across the globe indicates that the terrestrial biosphere is already experiencing a great planetary greening, likely in large measure due to the approximately 45% increase in atmospheric CO_2 since the beginning of the Industrial Revolution (Idso 2012).

In this regard, and with respect to *all* land plants, satellite-based studies consistently reveal net terrestrial primary productivity has increased by 6 to 13% since the 1980s. Additionally, Tans (2009) has reported that up until around 1940, Earth's land surfaces were a net source of CO_2-carbon to the atmosphere. From 1940 onward, however, the terrestrial biosphere has become, in the mean, an increasingly greater *sink* for CO_2-carbon. Other research from Ballantyne et al. (2012) shows the annual global net carbon uptake over the lands and oceans has doubled during the past five decades from 2.4 ± 0.8 billion tons in 1960 to 5.0 ± 0.9 billion tons in 2010.

What makes the above observations appear even more astonishing, however, is the fact that they have occurred despite many recorded assaults of both society and nature alike on planetary vegetation over this time period, including fires, disease, pest outbreaks, deforestation, and climatic changes in temperature and precipitation. Indeed, global temperatures significantly increased over the past century, rising to levels claimed to be unprecedented over thousands of years. Over just the past three decades, in fact, the Earth has experienced the warmest temperatures of the instrumental temperature record and a handful of intense and persistent El Niño events. That the biosphere experienced *any* productivity improvement over this time period, let alone a *doubling*, is truly amazing, and it demonstrates in large measure the powerful positive impact rising CO_2 is exerting on global vegetation.

Concluding remarks

There are two major viewpoints with respect to the potential impacts of a rising atmospheric CO_2 concentration and warmer temperatures on the productivity of Earth's biosphere. One, which is championed by climate alarmists and based primarily on computer-model projections, posits that increases in CO_2 and global temperature will have devastating effects on terrestrial vegetation, potentially causing the extinction of numerous plant species. The second offers an alternative perspective, postulating that the 'twin evils' of the climate alarmist movement (rising temperatures and atmospheric CO_2 concentrations) will lead to significant increases in the growth and productivity of Earth's plants nearly everywhere they are found. So, which hypothesis is correct?

Clearly, the evidence to date indicates that the ongoing rise in the air's CO_2 content is benefiting the biosphere. It is increasing the productivity of herbaceous, woody, and aquatic plants, improving their water-use efficiencies, and helping to ameliorate the many growth-retarding influences of various environmental stresses they face in the real world. The future of agriculture is bright, as yields are expected to increase, on average, by about one-third for a 300-ppm enrichment of the air's CO_2 concentration. Natural ecosystems are also expected to fare well. Forest growth rates, in particular, should continue their dramatic rise, and the ranges of several woody species will likely continue to expand. The amount of carbon that is removed from the atmosphere, incorporated into plant tissues, and ultimately sequestered in the soil should also increase.

Consequently, and in consideration of these benefits, the ongoing rise in atmospheric CO_2 should be widely celebrated. Carbon dioxide is not a *pollutant*, it is the very *elixir of life*.

14 The Poor Are Carrying the Cost of Today's Climate Policy

Dr Matt Ridley

Here is a simple fact about the world today:

- climate change is doing more good than harm.

Here is another fact:

- climate change policy is doing more harm than good.

These are both well-established facts, supported by a great deal of data, as I will demonstrate. Do these facts surprise you? It's certainly not the impression most politicians, scientists or journalists give. Yet the well-informed ones would not deny it if pressed. They would merely insist, instead, that this position will reverse later in the 21st century and that by then climate change, unchecked, will be doing more harm than climate policy. The eventual ends will begin to justify the painful means.

They may be right; we will see. But, today, we are deliberately causing suffering in partly futile efforts to stop something that is currently doing more good than harm, mostly to poor people.

And that should give us pause, at the very least. Is it right to ask today's poorest people – on whom the pain of climate policies fall most heavily – to make sacrifices for the sake of tomorrow's probably much richer people? Yet even to ask this question is to run a gauntlet of abuse from people, mostly paid by taxpayers, who accuse you of moral failings.

On no other topic that I write about do I get such vitriol and bitter criticism of my morality. When I made the argument on television once that climate change policy was hurting the poor, a prominent and wealthy left-wing commentator replied, 'But what about my grandchildren?' I am genuinely baffled as to why is it considered virtuous to cause pain to poor people today, and reward rich people, for the sake of the rich people's perhaps-even-richer grandchildren.

Eugenicists and population control advocates, incidentally, have made the same argument; we must harden our hearts and do painful things today for the sake of posterity.

During the great Irish famine, Charles Trevelyan, the Assistant Secretary to the Treasury in London, who had been a pupil of Malthus, called starvation an 'effective mechanism for reducing surplus population', adding: 'Supreme Wisdom has educed permanent good out of transient evil.'

In 1912 Leonard Darwin, son of Charles, argued, 'if wide-spread eugenic reforms are not adopted during the next hundred years or so, our Western Civilization is inevitably destined to such a slow and gradual decay as that which has been experienced in the past by every great ancient civilization'.

The ecologist Paul Ehrlich is an unabashed advocate of coercion to achieve population control, having said that to achieve it 'the operation will demand many apparently brutal and heartless decisions. The pain may be intense.' He called this 'coercion in a good cause'.

California's forced sterilisation programs in the 1920s, Germany's mass murders in the 1940s, India's semi-compulsory sterilisations in the 1960s, and China's one-child policy in the 1980s all justified huge suffering on the grounds that they would benefit future generations. Yet the demographic transition showed that the best way to reduce population growth is to be kind, not cruel; once babies survive, people plan smaller families.

My argument is not to be confused with the claim that climate change is not happening. Of course it is. Nor with the claim that it is all natural; I think it is highly likely that the increase in concentration of carbon dioxide (CO_2) over the past half century from an average of about 0.03% to an average of 0.04% of the atmosphere, small though it is, has had a warming effect. I am a card-carrying member of the overwhelming consensus that climate change is real and partly man-made. I also concede that climate change probably does already cause some harm in some places. The point is rather that the harm is currently smaller than the good it is doing, through longer growing seasons, milder winters, slightly higher rainfall, and faster growth rates of crops and forests because of CO_2 fertilisation. And that net good stands in stark contrast to the net harm caused by climate change policy.

The biggest way in which CO_2 emissions do good is through global greening. Ranga Myneni and colleagues (Zaichun et al. 2016) recently published evidence derived from satellite data showing that 25 to 50% of the vegetated parts of the planet has grown greener and just 4% browner, and that 70% of the greening can be attributed to an increased level of CO_2. The overall increase in green vegetation, which has occurred in all kinds of habitats – from the tropics to the Arctic, from deserts to farmland – is now estimated to be 14% during the last 30 years. This startling fact is confirmed by multiple other lines of evidence: tree growth rates; free-air concentration experiments in which the CO_2 level is enhanced over crops and natural habitats; increases in the amplitude of the CO_2 changes in the Northern Hemisphere each year; and so on.

Dr Zaichun Zhu from Peking University, the lead author of the Myneni paper (2016), described these results as follows: 'The greening over the past 33 years reported in this study is equivalent to adding a green continent about two times the size of mainland USA (18 million km²), and has the ability to fundamentally change the cycling of water and carbon in the climate system.'

Now just imagine if Zhu and Myneni had discovered the opposite: a 14% reduction in overall plant productivity over 30 years with browning in 37% of pixels and greening in only 4%. Most politicians, scientists, and journalists would have been screaming about it from the rooftops as an example of the harm caused by climate change. Behold the inherent bias towards suppressing good news that has plagued all debates about the environment for the past half century, and which has systematically misled the public. As I have consistently argued for years, the failure of doomsday predictions again and again is highly relevant data in this debate. But it is routinely ignored.

But back to global greening: if there is 14% more vegetation on the planet than 30 years ago, and 70% of this can be directly attributed to CO_2 fertilisation, this means there is more food for humans, animals, microbes, and fungi, and less land is needed to grow human food. There-fore, more land is available for nature than would otherwise have been the case if we had not raised CO_2 levels. It means richer biodiversity and less drought – as shown in Figure 14.1. It means lower food prices and less starvation. It means richer rainforests and less desert.

Figure 14.1 Fraction of global land in drought, 1982 to 2012

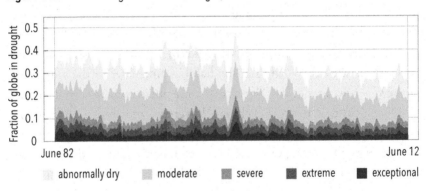

Standardized Precipitation Index data derived from MERRA-Land.

Source: Hao, Z et al., 2017, 'Global integrated drought monitoring and prediction system', *Nature*, viewed 13 May 2017, https://www.nature.com/articles/sdata20141. Reprinted under a Creative Commons CC-BY license.

For many years, Dr Craig Idso has been quietly and systematically collating the evidence as to how much faster crops have grown as a result of the CO_2 increase, detailing the many experiments that show very clearly that a higher CO_2 level makes plants more resistant to droughts – because they need to open their pores less and they lose less water as a result (Idso & Idso 2011). He has estimated the increased value of the world's crops as a result of the CO_2 fertilisation effect – over 30 years this increase comes to US$3.2 trillion (Idso 2013). That's $3,200,000,000,000.

Now if you argue that coal producers – like the one operating on my family's land, so, yes, I declare an interest – should be paying recompense for the damage they have done the world, you must also admit that they can take into account any benefit they have done. It's the net cost that counts. At the moment, it is mathematically indisputable that farmers owe coal producers a huge sum for supplying them with free CO_2 fertiliser. The burning of fossil fuels has boosted farm yields.

Incidentally, the CO_2 fertilisation effect seems to be working in the sea as well as on land. Some studies of eelgrasses (Palacios & Zimmerman 2007), seaweeds, other marine algae, and also corals (D'Olivio et al. 2013) indicate a positive effect from CO_2 enrichment. So do some studies of phytoplankton, which are responsible for much marine photosynthesis.

One laboratory experiment grew two strains of a diatom species and a coccolithophore species at 390 ppm and 750 ppm and found that 'increased CO_2 led to increased growth rates in all three strains ... enhancing growth rates 20%–40%.' They concluded that 'there could be a net increase in capacity for primary productivity at 750 ppm of CO_2, at least with regard to small diatoms and coccolithophores in coastal environments' (McCarthy et al. 2012).

Furthermore, numerous studies suggest that the slightly lower pH of seawater resulting from CO_2 enrichment does not seem to adversely affect growth rates in such calcifying phytoplankton. For example, one study concluded that 'the coccolithophore, *E. huxleyi*, has an ability to respond positively to acidification with CO_2 enrichment'. This 'suggests that physiological activities of *E. huxleyi* cells will not be seriously damaged by ocean acidification at least up to 1200 ppm CO_2 in the atmosphere' (Fukuda et al. 2014). Another study concluded that 'carbonate chemistry is not the sole and overriding control over coccolithophore calcification', and that this should, 'seriously call into question' the notion that 'ocean acidification will lead to a replacement of heavily calcified coccolithophores by lightly-calcified ones' (Smith et al. 2012).

Laboratory experiments using a common reef-building coral found that its growth rate increased as the CO_2 level was raised and remained high even with a CO_2 level of 0.06%, that is to say, half again as high as today's.

Global warming itself has benefits for people, too. They include: fewer winter deaths; lower energy costs; better agricultural yields; probably fewer droughts because of increased rainfall; and maybe richer biodiversity. These turn out to be real and large effects.

Climate change policies

The policies designed to slow global warming, meanwhile, have huge costs. Let me walk you through the details in case you are doubtful. Here are ten examples of the harm done by policies designed to solve the problem of climate change.

1. Ethanol subsidies

Ethanol subsidies in the United States, Brazil and in Europe were introduced specifically and explicitly to reduce CO_2 emissions. Yet they did environmental harm. As *Bloomberg* (2016) reports:

> The Natural Resources Defense Council used a 96-page report in 2004 to proclaim boundless biofuel benefits: slashed global warming emissions, improved air quality and more wildlife habitat. Instead, farmers ploughed

millions of acres of prairie grasses to grow corn for making ethanol, with fertilizer runoff contributing to a dead zone in the Gulf of Mexico. Scientists warned that carbon dioxide emissions associated with corn-based ethanol were higher than expected.

And they did very little if anything to reduce emissions.

Ethanol has now displaced a bit more than 0.5% of world oil use. This ethanol conversion consumes about 5% of the world's grain crop, which in turn raises food prices. The United Nations (UN) Food and Agriculture Organization produced a report concluding it was one of the main reasons that the price of food shot up in 2008, and stayed high for some years afterwards until harvests began to catch up, worsening malnutrition and starvation, and encouraging the destruction of rainforest to cultivate more land. The policy of ethanol subsidies steals land from nature. Worse, it steals food from poor people and puts it in rich people's cars.

Indur Goklany has estimated that this policy kills almost 200,000 people a year (Bryce 2010). As a farm owner, I probably benefit a little bit from these programmes, so I have no vested interest in criticising them. But that's not going to stop me.

2. Biodiesel programmes

Biodiesel programmes for making motor fuel from palm oil in the tropics, and from rapeseed oil in Europe, are all subsidised by the European Union specifically to reduce emissions, but they actually increase them. Transport & Environment, a green group, has calculated that by 2020, biodiesel will increase emissions from transport by 4% compared with using fossil-diesel, equivalent to putting an extra 12 million cars to the road (Gosden 2016). The subsidies also encourage the destruction of rainforest and the cultivation of land that would otherwise be available to nature. As a farm owner, I probably benefit slightly from this harmful policy.

3. The promotion of diesel cars

Europe mandates the promotion of diesel cars through the tax system as a deliberate policy to reduce CO_2 emissions. Diesel engines have lower CO_2 emissions than petrol engines, but higher emissions of nitrogen oxides and particulate emissions, which are more dangerous.

A study by Steve Yim and Steve Barrett (2012) determined that there were nearly 5000 deaths each year caused by these type of vehicle emissions. This policy, therefore, exacerbated this issue, contributing to these deaths. The scandal only came to light when it emerged that Volkswagen and other car manufacturers cheated on the emissions tests. As a diesel car driver, I benefited from this lethal policy.

4. Burning pellets derived from wood products

Britain is now burning pellets derived from wood products in power stations to produce electricity. The pellets, euphemistically called 'biomass', are harvested from forests in South Carolina and other parts of the United States (Ernstig 2015). Contrary to popular myth, these are derived not from wood by-products of a harvest primarily taken for other purposes, but from roundwood.[1] Wood residues are used to dry the biomass prior to pelletisation (Stephenson & MacKay 2014; Rose 2015).

Burning wood produces more CO_2 than burning coal, for every unit of energy generated. It also encourages deforestation and habitat destruction, raises the price of electricity for consumers, and requires shipping combustible material a third of the way around the world. That wood regrows but coal does not is of little comfort given that wood takes several decades to regrow. As a landowner, this policy helps me; the higher price of wood has helped reduce my losses on managing woodland. But it also hurts me because I make money from coal.

---

1   Timber left over as small logs.

5. Wind power

Many countries have littered their rural beauty spots with 120-metre towers of steel, standing in massive reinforced concrete bases and equipped with electrical dynamos, whose two-ton magnets are about 50% rare-earth metals – usually neodymium, which is mined and refined in a very dirty process in China with toxic and radioactive waste as a by-product. These windmills produce expensive, unreliable, and intermittent electricity far from where and when it is needed, requiring expensive back-up power, and costly and unsightly power lines. They are subject to huge subsidies, which go mostly to the rich, including landowners like me, and which hit the poor harder than taxes would do because the money for these subsidies are levied on electricity bills. Indeed, I get some income from a wind turbine built on land I do not own but for which I have the mineral rights, beneath. They kill thousands of rare birds of prey, gannets, swifts, and other soaring birds, as well as large numbers of bats. And they do very little to reduce emissions (Fisher & Fitsimmons 2013; Hughes 2012; Ridley 2012).

6. Solar farms

In cloudy Britain and Germany very large sums have been diverted from relatively poor people to wealthy landowners to cover good agricultural land with solar farms, which produce very little electricity, and do so mainly when it is least needed on warm summer afternoons. These solar farms shade the ground preventing the growth of plants that could feed either people or wildlife. They also mean less land is available for nature. Their contribution is trivial; to the nearest whole number, solar power still produces zero percent of global energy needs. As a landowner, I have been offered large annual incomes for installing solar plants. To the consternation of my accountant, I have refused them. At last a climate policy that has not benefited me.

7. Renewables, only

Western governments, the International Monetary Fund, and the World Bank have all said that to avoid increasing emissions they are no longer willing to grant aid for the building of fossil-fuel plants in the poorest countries. They prefer to spend the money on renewables instead, where it goes less than half as far. The result is that the death toll of those who die as a result of cooking over open wood and dung fires for lack of electricity – currently about three million people a year – will not fall as fast as it should, while millions more are missing out on the benefits of electricity, including refrigeration and education.

The Center for Global Development has estimated that US$10 billion invested in gas-fired generation in sub-Saharan Africa would meet the needs of 90 million people. The same sum spent on renewable energy technology would help just 27 million people (Moss & Leo 2014). The overall conclusion from this study was that more than 60 million additional people in poor nations could gain access to electricity if investment were allowed in natural gas projects, not just renewables.

8. Fuel poverty

The effect of renewable energy is to drive many poorer people into fuel poverty, so that they struggle to stay warm in cold winters. Today, throughout the world, far more people die of cold than of heat. One study (Public Health England 2014) concluded that winter deaths exceeded summer deaths in all 31 European countries, on average by 14%, and the total excess winter deaths between 2002–03 and 2010–11 was more than two million. On average, 65 British people a day are dying because they cannot afford to heat their homes properly. So climate change helps to reduce winter mortality; climate change policy helps to increase it again.

9. High energy costs

The high energy costs resulting from climate policies have resulted in the closure of heavy-industrial plants throughout Britain and other parts of western Europe, throwing many thousands out of work (Montford 2015). Few emissions savings have resulted, however, because many of the jobs have simply moved to China and India. The total cost of Britain's climate policies during the 21st century is on course to reach £1.8 trillion. The total benefit from that spending is expected to be a lowering of the average temperature by about 0.005 °C — that's half of one-hundredth of a degree. And for that we are destroying industry.

10. The neglect of more serious environmental problems

The immense diversion of political energy and finance into studying and mitigating climate change has starved attention from far more imminent, serious and soluble environmental problems, such as invasive plant and animal pests, habitat loss, and the overfishing of the oceans. These are neglected and underfunded because so much money and prestige is spent on climate change.

Conclusion

I could go on, but you get the point. Climate policies really hurt people and the planet.

Notice, by the way, that the beneficiaries of these policies are mostly comparatively rich people: landowners, investors, scientists, policy advisers, the employees of non-profit organisations (NGO). The victims are mostly poor people: subsistence farmers, poorer pensioners, manual workers. None of this is controversial, let alone imaginary. All these effects are real. Environmentalists concede that these things are happening. Some claim that they are justified because they will avert future disasters. Indeed, some claim such disasters are already upon us.

But are they? There has been no increase in extreme weather: no trend towards stronger or more frequent storms, no consistent change in the occurrence or intensity of tornadoes or cyclones or lesser storms. The IPCC has confirmed this again and again. The cost of storm damage has increased, but this is entirely due to the increase in the value of property, not the worsening of storms. Indeed, as a proportion of gross domestic product (GDP) storm damage has been falling, not rising. Flooding is worse in many parts of the world, but largely because of land-use changes – usually deforestation or drainage – causing run-off to be more rapid.

As for droughts, there has been, if anything, a very slight *decline* in the frequency and severity of drought over recent decades, while famine has largely vanished from the face of the Earth for the first time in recorded history, except under a few autocratic regimes like that of North Korea. The vast famines that plagued the twentieth century are not happening in the 21st.

Again, these facts have been confirmed by the IPCC itself.

In its latest report, it backs off claims made in its previous one:

> The most recent and most comprehensive analyses of river runoff do not support the IPCC Fourth Assessment Report (AR4) conclusion that global runoff has increased during the 20th century. New results also indicate that the AR4 conclusions regarding global increasing trends in droughts since the 1970s are no longer supported. There is low confidence in a global-scale observed trend in drought or dryness (lack of rainfall), owing to lack of direct observations, dependencies of inferred trends on the index choice and geographical inconsistencies in the trends.

As far as we can tell – the interpretation of global sea-level data is not straightforward because of changes in techniques and adjustment for local tectonic factors – sea level has not risen much faster, if at all, in the past three decades than it did in most of the past century. Satellites suggest it is currently going up at about 34 cm a century – which is about 3.4 mm a year.[2] Small, uninhabited islands are sometimes lost to the sea mainly because of tectonic sinking rather than sea-level change. A study

---

2   http://climate.nasa.gov/vital-signs/sea-level/.

by Webb and Kench (2010) of 27 coral atolls in the central Pacific over several decades found that more atolls increased in size than decreased, despite rising sea levels. As they stated, it is in the nature of atolls to rise with sea level through coral growth and sand accumulation: 'Islands are geomorphologically persistent features on atoll reef platforms and can increase in island area despite sea-level change' (Webb & Kench 2010).

Arctic sea ice has declined in summer, but Antarctic sea ice has increased, and the decline in Arctic sea ice has had no measurable deleterious effect on either people or polar bears. Indeed, the trend in polar bear numbers is not down and may be upward as they recover from past hunting.[3] Most glaciers have retreated, as they have since about 1850 (that is, before man-made climate change), but again without significant impact on human welfare.

So, it is very clear that climate change has done very little harm so far and is doing very little harm today. On balance it has done net good. As I explained in the beginning, most well-informed politicians, journalists and scientists accept that this is the case. So, does the IPCC. They all say the damage will nearly all be in the future. So I am not saying anything controversial here, let alone outside the consensus.

Studies of the 'social cost of carbon' and of the economic impacts of climate change on average find that climate change is not yet doing net harm, and will only do so when the temperature reaches about 2 °C above pre-industrial levels. The IPCC confirmed this in its latest assessment report. The opening words of the executive summary of Chapter 10 of Working Group 2's report reads (Field et al. eds. 2014):

> For most economic sectors, the impact of climate change will be small relative to the impacts of other drivers (*medium evidence, high agreement*). Changes in population, age, income, technology, relative prices, lifestyle, regulation, governance, and many other aspects of socioeconomic

---

3   http://www.iucnredlist.org/details/22823/0.

development will have an impact on the supply and demand of economic goods and services that is large relative to the impact of climate change.

Globally, climate change policy is doing harm, while climate change is doing good. Locally, in the poorest countries, the effect can be even more stark. Here the pain of policy is most acute and the gain of changing the concentration of atmospheric CO_2 is most dramatic.

Take Niger, the fourth poorest country in the world with a per capita GDP about 1% of Britain's. One reason for that poverty is the very low level of energy available to Niger's people, final energy consumption being one of the lowest in the world. The small amount of electricity available to a small proportion of the population comes from one tiny coal-fired power station, some diesel power stations, and an interconnector from Nigeria. About 90% of Niger's households depend on wood for cooking – which means appalling levels of premature death from smoke inhalation. The government has this to say:

> Butane gas or LPG, a fuel currently available in sufficient quantity in Niger (44,000 tons per year) and which should be the solution to replace firewood as cooking fuel used by households, requires the acquisition of accessories for its use. These are accessories that, although available on the market, are not within the reach of low-income households and thus many in Niger have no other choice than to make use of traditional energy sources (Gado 2015).

So, thousands of lives could be saved, and living standards raised if the aid money could be used to buy butane stoves. That would stop the continuing devastation of Niger's forests and scrublands, too. Instead, many westerners insist that Niger must go green: 'Niger needs to develop an energy policy that embraces renewables as part of a longer-term energy vision,' says the International Renewable Energy Agency (2013). Wood is, of course, renewable, but that is not what they mean. Wind and solar power are dreadfully expensive, as well as unreliable, and

Niger cannot afford them. The refusal of the West to support fossil fuels in such countries kills people and damages the environment.

Niger has one thing in its favour. It is smack in the middle of the Sahel, the region that has seen the fastest global greening. Because of the increase in CO_2 in the air, as a result of the burning of fossil fuels, plants in this semi-arid area can now grow faster and lose less water from the pores in their leaves as they do so: water efficiency improves. An increase in rainfall in the Sahel is also a result of global warming.

'Recent trends show good signs on the recovery of the region, and the relative vegetation index (NDVI) presented for the country of Niger in this map shows increases in the period from 1982 to 1999,' reads a recent report (Ahlenius 2006). Some imaginative land-management policies have helped. So Niger has seen an increase in woody vegetation, even as its people desperately chop away at it to provide themselves with cooking fuel.

Niger is a perfect example of the horrible hypocrisy of the climate establishment. Niger's extreme poverty makes it vulnerable to the effects of climate policy; but it currently benefits from climate change itself.

I say to my greener friends:

- Where do you get your insouciance about the clear evidence that the poorest people in the world are the ones hardest hit by climate change policy today?
- Where do you get your indomitable certainty that the end justifies these means?
- Where do you find the evidence that we must cause certain pain to today's poor in order to forestall the small possibility of suffering among tomorrow's rich?
- And where do you find the hubris to occupy the moral high ground?

15 The Impact and Cost of the 2015 Paris Climate Summit, with a Focus on US Policies

Dr Bjørn Lomborg[1]

Introduction

Global warming is a real phenomenon, it is mostly man-made, and it will have a long-run overall negative impact. This says nothing, however, of the efficacy of government attempts to reduce temperature through international agreements, nor of the economic costs involved in implementing such promises. As such, my main aim in this chapter is to estimate both the *temperature impact* and the *economic cost* of the promises associated with the 2015 Paris Climate Summit (henceforth, Paris), with a special focus on the promises of the United States. What I find is that these promises will do little to tackle the problem of global warming (that is, they will have negligible impact), and that implementing the promises will cost at least nearly $1 trillion a year in foregone gross domestic product (GDP) by the year 2030.

At the time of writing, mine is the only peer-reviewed estimate of Paris, and it shows that adopting all promises from 2016–2030 will reduce the temperature increase in 2100 by 0.05 °C. Further, even if we optimistically assume all countries continue their promised reductions

---

1 This chapter was adapted from testimony delivered by Bjorn Lomborg on Tuesday, 1 December 2015, at the US House of Representatives Committee on Science, Space and Technology.

until 2100, the temperature increase would be reduced by just 0.17 °C. Focusing specifically on the US's policies also demonstrates a negligible impact. For instance, by implementing the US Clean Power Plan global temperatures are estimated to reduce by maximally 0.013 °C in 2100. Even if the US further implements the entire set of US climate promises (not just the US Clean Power Plan) it will reduce global temperatures in 2100 by just 0.031 °C.

By using the best available climate-economic model ensembles – from the Stanford University Energy Modeling Forum (EMF), the Asia Modeling Exercise (AME), and the European Union (EU) and US Environmental Protection Agency (EPA) CLIMACAP-LAMP[2] project – we can estimate the economic cost of implementing the Paris promises. The foregone economic output from adopting and implementing the Paris promises is approximately $924 billion in GDP per year by 2030. And that's only if the policies are employed efficiently and effectively. If the climate policies are not employed efficiently, then the peer-reviewed literature suggests the *economic costs will likely double to almost $2 trillion per year.*

The cost for the US climate promises alone is likely to range from $154 billion to $172 billion every year in lost GDP by 2025, and double that if not enacted efficiently.

Which promises should be included in a Paris analysis?

When deciding which promises should be included in my analysis, I explicitly limit these to the promises of policies that have practical political implications soon, and have a verifiable outcome by 2030. That is, I do not incorporate policies that merely promise actions starting only or mostly after 2030.

---

2   The Integrated Climate Modelling and Capacity Building Project (CLIMACAP) working with the Latin American Modeling Project (LAMP), is financed by the European Commission (EU) and the US Environmental Protection Agency (EPA).

Some critics claim that in analysing Paris we should look not just at the promises made for the years 2016–2030, but also at those promises that are much further out, or into the future, such as the Chinese promise to reduce emissions after 2030, the US's promise to reduce its emissions to 80% below the 2005-level by 2050, and the EU's promise to cut emissions to 80–95% below the 1990-level by 2050. Such an interpretation is implausible for three, interlocking, reasons.

First, *it is difficult to defend the inclusion of targets with a very low likelihood of implementation.* I only include policies that have practical political implications soon and have a verifiable outcome by 2030. It is undeniable that political targets further into the future are less likely to be implemented. Recent history clearly indicates that climate promises even 10 to 15 years ahead will be routinely flouted.

When China commits to reduce its carbon intensity of GDP by 60–65% below the 2005-level by 2030, we can analyse the progression towards that goal very clearly over the next 15 years, and clearly determine if it is met by 2030. As such this is included within my analysis. However, the promise to 'achieve the peaking of carbon dioxide [CO_2] emissions around 2030 and making best efforts to peak early' – often curiously misquoted, as, for example, 'peak CO_2 emissions by 2030 at the latest' (Climate Action Tracker 2016a) – is something that will only have an effect *after* about 2030, and it is something that will first be verifiable in about 2035, or later.

This is especially true given that Chinese energy statistics are notoriously opaque. Just recently it became clear that China burned perhaps 17% more coal per year in recent years than was previously understood (Buckley 2015). China's 'peaking' promise is very unlikely to be achieved based on economic reality alone. The cost can be identified from the AME, which indicates that the lowest GDP loss would be about US$400 billion, or about 1.7% of GDP, and likely twice that. It strains credibility to expect China to commit such economic self-harm. It is worth noting in passing that in its INDC China also promises to be 'democratic' by 2050 (China INDC 2015). The one-party state's vow should probably be treated rather similarly to the suggestion that it will rein in economic growth so dramatically.

Second, my approach is *methodologically clear.* The alternative is unable to avoid a slippery slope that would include every target, vow, promise, or vague political undertaking. In my analysis, I was consistent in ruling out longer-term promises that were further into the future and economically implausible. Adopting a tight definition avoids an obvious slippery slope towards a ridiculous premise that, since almost all states have already accepted the 2 °C promise, if all promises are included, then by default we will see temperatures rise less than 2 °C. Including these promises would make a mockery of any real analysis of what the Paris treaty can achieve.

Indeed, since almost every nation has signed up to reduce temperature rises to 2 °C, and about 80 to 90 nations including the EU and the US 'endorse' (United Nations 2010) this target in their INDCs, where should we draw the line? If we were to include the Chinese 'peaking' promise, why not also include the US promise to cut 80% by 2050 and the EU promise to cut 80–95% by 2050, both of which are mentioned in their INDCs? As such, I also left out the US promise of 'deep, economy-wide emission reductions of 80% or more by 2050'. And I left out the EU promise 'to reduce its emissions by 80–95 per cent by 2050 compared to 1990'. Data from the Stanford EMF shows the average GDP loss at almost €3 trillion annually, if done efficiently. If not, the cost will likely double to almost €6 trillion or 25% of EU GDP in 2050 (Knopf et al. 2013).

Third, *the commitment period of 2016–2030 is by far the most common understanding of what Paris constitutes.* This is true whether we pay attention to the United Nations (UN) or to the official material from nations themselves. The United Nations Framework Convention on Climate Change (UNFCCC) itself describes the central results as emission reductions achieved in 2025 and 2030, not further. It specifically labels possible emission reductions after 2030 as actions taken by nations 'beyond the time frames stated in their INDCs (e.g. beyond 2025 and 2030)'. The US clearly states that its understanding of its INDC is for 2025 and not further: 'The U.S. target is for a single year: 2025' (US INDC 2015). The EU sets its targets for 2030 and not any further. In its own INDC, China clearly writes what it expects from the Paris agreement, namely to 'formulate and implement programs and measures to reduce or limit greenhouse-gas emissions for the period 2020–2030' (China INDC 2015). So even China itself is unequivocal that the Paris deal is not about promises *after* 2030, but, instead, *up until* 2030.

It is also worth remembering that previous promises have routinely been flouted, which lends less credibility to new promises, especially those which are far off. Consider an analysis conducted in 1997 on the likely effect of the Kyoto Protocol. Should it have included not just the specific commitments made in Kyoto, but every far-reaching promise made around that time? Likely not, because we could have assumed that not only would this treaty be implemented, but that stronger and ever-increasing cuts would consistently be made as a result of policy (and not economic downturns) for decades. History shows that we would have been utterly wrong to do so.

Moreover, should such an analysis of Kyoto have included President Bill Clinton's (1993) announcement that the US would reduce its emissions by 2000? That promise was never fulfilled. According to the *Washington Post*, the US administration's excuse was that the 'goal is no longer possible because the economy has grown more rapidly than expected' (Joby & Baker 1997). The commitment failed, even though it was for just seven years later, it was to be implemented right away, and it was going to be implemented under the same president who made it.

In 1992, every industrialised nation promised to return their emissions to the 1990-level by 2000 (United Nations 1992) – and almost every single OECD country missed that target. Even the commitments made in the Kyoto Protocol itself ended up meaning nothing. The treaty was abandoned by the US, and eventually by Russia, Japan and Canada. We would clearly not have known this if we were conducting an analysis in 1997; however, these examples show that it is folly to assume that we can realistically believe targets much further into the future to be accurate.

The temperature impact of the 2015 Paris Climate Summit (Paris)

What do we already know about the temperature impact of Paris? At the time of writing, there is only one peer-reviewed paper estimating the temperature impact of the promises made in the Paris Climate Summit. That paper is my paper (Lomborg 2016). In it I investigate the *change* in temperature in 2100 from implementing promises for Paris using the climate model MAGICC 6.3. This is the latest version of a simple climate model used in all the five IPCC assessment reports from 1990 to 2014 (Edenhofer et al. 2014; IPCC 2007; United Nations 2006).

All the following runs use default values of MAGICC with a climate sensitivity of 3 °C. Sensitivity analysis shows that different models and carbon cycling does not substantially change the outcome. Two standard climate scenarios are run (RCP8.5 and RCP6) both with and without the promised reductions. Because sensitivity analysis shows that the outcome does not substantially change, only RCP8.5 is demonstrated below. In the following section I begin with the temperature impact of the US promises, before examining the global impact.

United States impact

We can begin by specifically examining the US Clean Power Plan, which is a subset of the broader US climate policies. Published on 3 August 2015, the US Clean Power Plan requires the US power sector to reduce emissions by 535 Mt (million metric tons) of CO_2 every year by 2030, compared to the expected emissions from the US power sector, as estimated by the Energy Information Agency (EIA) (2016).

When examining political promises, it is useful to examine multiple scenarios for future policy changes. In my analysis, I take two different scenarios: 'optimistic' and 'pessimistic'. The real outcome, of course, is likely to be somewhere between the optimistic and pessimistic scenarios.

The optimistic scenario is where the US will continue their Clean Power Plan forever. That means the US emissions will forever be 535 Mt CO_2 lower than the baseline as given by the EIA. The pessimistic scenario is where the US will live up to its Clean Power Plan promises by 2030, but then fall back to its baseline emissions as estimated by EIA. Specifically, this scenario is modelled as halving the emission reduction target every decade, on the understanding that the system will trend towards higher emissions as the restrictions are lifted.

Running the MAGICC climate model with the RCP8.5 global emissions – both with and without the reductions in CO_2 emissions from the US Clean Power Plan shows the practically negligible effect of these policies on temperature. The temperature reduction resulting from the pessimistic scenario of the US Clean Power Plan – that is where the policy is gradually abandoned after 2030 – will reduce global temperatures by 2100 by 0.004 °C. In the more optimistic scenario – that is where the US Clean Power Plan is continued throughout the next century – reducing 535 Mt CO_2 each year from the baseline, will reduce global temperatures by 0.013 °C.

Following from the promises above, the US administration has also promised in its 'intended nationally determined contribution' (henceforth, US INDC) for Paris that it will reduce its overall greenhouse-gas emissions to 26 to 28% below the 2005-level by 2025 (US INDC 2015). The US is very clear that this is a one-point target for a single year: 2025. This reduction promise works out to about 1.27 Gt

CO_2 equivalents (e)[3] in 2025. Given that this is delivered faster and is also more than double the US Clean Power Plan outlined above, the temperature impact, as demonstrated below, is similarly more than twice as large.

As above, I examine two possible scenarios. The first pessimistic scenario is where the US will only live up to the letter of its US INDC promise, cutting 1.27 Gt CO_2e in 2025, but then reverts back to the baseline. The optimistic scenario sees the US living up to its promise not just in 2025 but every year thereafter, reducing 1.27 Gt CO_2e from the baseline throughout the 21st century. Running the MAGICC climate model with the RCP8.5 global emissions – both with and without the reductions in CO_2 emissions from the US INDC promises – achieves the output below.

The temperature reduction resulting from the entire US Paris promises by the end of the century will be only 0.008 °C if the policy is gradually abandoned after 2025 (the pessimistic scenario). If the US Paris promise is continued throughout the century as the optimistic scenario it will reduce global temperatures by 0.031 °C by 2100.

Since these estimates include the US Clean Power Plan (as outlined in the previous section) the *net effect* of the extra US Paris promise is, in the optimistic case, 0.018 °C.

Global impact

What about the other global players and their contribution to the Paris emission reductions?

The US, China and the EU reductions approach almost 80% of the total promised reductions, and as such I focus on the latter two, below.

In 2009, the EU legislated the climate policy promise of reducing greenhouse-gas emissions to 20% emission reductions below the

---

3   CO_2e, or carbon dioxide equivalent, is a standard unit to express the impact of all greenhouse gases in terms of the amount of CO_2 that would create the same amount of warming.

1990-level by 2020. This is the EU 20-20 policy. This leads to 0.93 Gt CO_2e annual reduction by 2030. In its INDC for Paris the EU promises a 40% reduction below the 1990-level by 2030, resulting in an annual 2.1 Gt CO_2e reduction by 2030. China has promised to reduce its CO_2 intensity to 60–65 per cent below the 2005-level, which translates into a 1.95 Gt CO_2e annual reduction promise by 2030.

I estimate that the rest of the world (RoW) INDCs will reduce emissions by 1.48 Gt CO_2e annually by 2030 (Boyd et al. 2015). That means, in total, the INDCs for Paris will result in an emission reduction of 6.8 Gt CO_2e by 2030.

Figure 15.1 Global temperature anomaly, 2000–2100

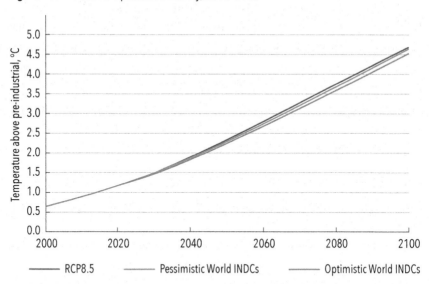

The temperature anomaly is calculated for the period 2000-2100 from baseline RCP8.5, showing both optimistic and pessimistic global Paris promises (Global INDCs), run on MAGICC.

Source: Republished under the terms of the Creative Commons Attribution-NonCommerical-NoDerivs License by permission from Durham University and John Wiley and Sons Ltd – Lomborg, B 2016, 'Impact of current climate proposals', *Global Policy*, vol. 7, no. 1, pp. 109-118, copyright 2015.

In Table 15.1, one can see the total impact of these policies in degrees Celsius. The total impact of all INDC climate policies will reduce temperatures in 2100 by 0.048 °C if the Paris promises are gradually abandoned after 2030. If the Paris promises are kept throughout the century, reducing emissions by 6.8 Gt CO_2e below the baseline every year, the result will be a more optimistic temperature reduction of 0.170 °C by 2100, as shown in Figure 15.1.

Table 15.1 Impact of climate policies on temperature

Reduction in temperature		
°C year 2100	Pessimistic	Optimistic
US INDC	0.008	0.031
US CPP	*0.004*	*0.013*
EU 2030 INDC	0.017	0.053
EU 2020	*0.007*	*0.026*
China INDC	0.014	0.048
RoW INDC.	0.009	0.036
Global	0.048	0.170

The impact is measured in terms of reduction in temperature by 2100 in °C, optimistic and pessimistic for RCP8.5, using MAGICC.

Source: Republished under the terms of the Creative Commons Attribution-NonCommerical-NoDerivs License by permission from Durham University and John Wiley and Sons Ltd – Lomborg, B 2016, 'Impact of current climate proposals', *Global Policy*, vol. 7, no. 1, pp. 109-118, copyright 2015.

The economic cost of the 2015 Paris Climate Summit (Paris)

While it is now clear that the impact of Paris on global temperatures is going to be difficult to measure, in this section I use existing data to make a reasonable first estimate of the total *economic cost*. Such an estimate is necessary because, extraordinarily, there seem to be no official estimates of the costs of the proposed Paris climate policies, either for the US, the EU, China, or for the entire world.

Europe's climate promises are probably the best documented in the peer-reviewed literature, but this literature also clearly shows that the studies typically lag political decisions by some years. Therefore, we have good estimates for previous decisions but much less exact estimates for the ones the world is thinking of committing to with Paris.

The Stanford EMF – the gold standard for the economics of climate and energy – has done several studies of the previous EU climate policy, which promised a 20% reduction in CO_2 emissions from the 1990-level by 2020. These studies showed two main things. First, and perhaps not surprisingly, in the rare cases where official cost estimates are made, they are *often much underestimated.* That is, governments make enormous underestimates of the economic costs of adopting climate policies.

For instance, the EU estimated that the total cost of its 2020 policies could be as little as an annual 0.4% of GDP loss – or €64 billion, annually (Capros et al. 2011). In stark contrast, the peer-reviewed cost came in more than three times larger, at 1.3% of GDP loss – or €209 billion, annually (Tol 2012). Similarly, the Mexican government assumed its climate policies would cost US$6 to $33 billion, annually, by 2050 (Veysey et al. 2016). The peer-reviewed literature – supported by the US Environmental Protection Agency (EPA) and the EU – shows that this is 'far lower than any of the cost metrics reported by the CLIMACAP-LAMP models'. Indeed, the cost in 2050 is found to be between 14 and 79 times higher than the government estimate (approximately US$475 billion, annually).

Second, the EMF studies show that *politicians rarely pick the most efficient climate policies.* Governments don't tend to implement policies that cut CO_2 for the lowest possible cost. Failure to pick efficient policies typically doubles the economic cost of intervention. For instance, the EU could have reduced its emissions by switching to gas and improving efficiency for a GDP loss of 0.7% (Böhringer et al. eds. 2009). However, phenomenally inefficient solar subsidies and biofuels are often more alluring, which is why the actual EU cost almost doubled to 1.3% of GDP.

In the following sections, I tally the costs for the US, EU, Mexico, and China, which together make up about 80% of the total promised reductions under Paris, thereby giving a good representation of their economic cost.

There is no official estimate for the cost of the US promise to cut 26 to 28% of its greenhouse gases by 2025. We can turn to the Stanford EMF for the US – the so-called EMF 24 – which has run more than 100 scenarios estimating all greenhouse-gas emissions and the GDP costs associated with those (Fawcett et al. 2013). Estimating the foregone GDP with a regression across all these data points suggests that cutting 26% of greenhouse gases by 2025 results in a GDP loss of approximately US$154 billion, annually, for the US. Further, a 28% greenhouse-gas reduction incurs an annual GDP loss of $172 billion for the US.

The EU promises to cut its emissions by 40% below the 1990-level by 2030 (EU INDC 2015). While there are no official estimates of the cost, the latest peer-reviewed Stanford EMF for the EU – the so-called EMF 28 – estimates costs from a number of different reductions (Knopf et al. 2013). The closest policy attempts to reduce emissions by 80% in 2050, leading to an average reduction of 41% by 2030. That reduction across the models that estimate GDP loss is equivalent to reducing EU's GDP by 1.6% GDP in 2030 – or €287 billion (or US$305 billion at the rate of exchange in 2010).

China has promised to reduce its energy intensity to at least 60% below 2005, which is equivalent to reducing its emissions by at least 1.9 Gt CO_2e each year (China INDC 2015). In the international research project – the AME – nine energy-economic models estimate what different efficient reduction policies will attain in emission reductions and GDP reductions (Calvin et al. 2012, Calvin et al. 2012). Using the AME data, it is likely that China can reduce 1.9 Gt CO_2e for about $200 billion in annual GDP loss.

Another well-documented cost is for Mexico, which has enacted the strongest climate legislation of any developing country, conditionally promising to reduce its emissions by 40% below what it would otherwise have emitted by 2030 (Climate Action Tracker 2016b). The cost estimates of the Mexican government, as mentioned in the previous section, are approximately 14 to 79 times lower than the actual cost estimated in a new study supported by the US EPA and the EU. The CLIMACAP-LAMP project has estimated costs throughout Latin America. This peer-reviewed analysis for Mexico finds that the Mexican cost in 2030 is about 4.5% of GDP or about US$80 billion, annually (Veysey et al. 2016).

Therefore, the total cost of US, EU, China and Mexico adds up to US$739 billion (or US$757 million, if the US goes for 28%). Given that the reductions from US, EU, China and Mexico add to about 80%, it is reasonable to assume that the US$739 billion constitutes 80% of the total cost, making the global cost approximately US$924 billion.

Table 15.2 Cost of Paris promises, GDP loss per year

Foregone GDP, $billions per year	Most effective policy	Most likely policy
USA	154	308
EU	305	610
China	200	400
Mexico	80	160
Rest of World	185	370
Global cost	**$924 billion**	**$1848 billion**

Effective policy is based on best multi-model estimates from EMF24, EMF28, AME and CLIMACAP-LAMP, and showing a comparative estimate for rest of the world. The 'most likely policy' column simply assumes costs to double, which is a likely scenario, as shown by the EU and other climate policies.

As illustrated by Table 15.2, it is estimated that there is a cost of US$924 billion in annual lost GDP by 2030 if all nations enact the most efficient climate policy (which is likely to be an increasing carbon tax that is uniform across sectors and countries). However, as elucidated in the previous section, and as previous experience shows, an effective climate policy formulation is very unlikely, implying that the total cost may double to a global cost of approximately US$1848 billion (Böhringer et al. 2009).

It is therefore likely that the global cost of Paris will reach at least US$1 trillion, annually, by 2030, while the cost associated with realistically less-efficient policies could very likely get close to US$2 trillion, annually. That is an enormous economic cost, especially given the small temperature impacts as previously outlined.

Considering other findings, and across the entire spectrum of estimations of Paris, there is a great deal of agreement. As is evident in Figure 15.2, all find that with the Paris promises, emissions will be around 53.7 to 57.6 Gt CO_2e in 2030. My value is within 0.2 Gt of the median.

However, if we look at the baseline – no climate policy through the 21st century – the expected cumulative emissions between 2010 and 2100 in Figure 15.3 has significant outliers on both sides. On the low end is MIT, which is likely because it has already included emission reductions from climate policies until 2014. Just above 7000 Gt CO_2e lies AME, Lomborg, EMF27, and UNEP. On the high end is CAT and Climate Interactive (CI) – with CI more than 2000 Gt higher.

The CI emissions are actually higher than any model in EMF27. That means CI can claim that Paris, or any other policy, will reduce about 2000+ Gt more emissions than any other analysis. Such a claim, however, is of course entirely spurious. Since the unrealistically high baseline is entirely made up, these emissions would never have taken place, and hence Paris climate promises can't take credit for eliminating them.

If we look at the cumulative emissions from the Paris INDCs in Figure 15.4, it is clear that MIT, CI and Lomborg find about the same

Figure 15.2 Global annual emissions estimated for 2030, all Paris promises implemented

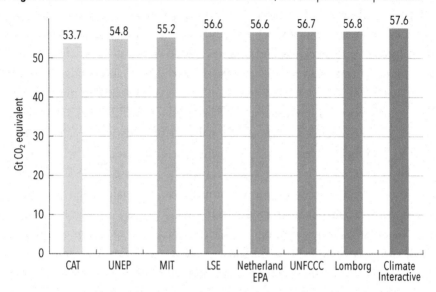

Global annual emissions estimated for 2030 with all INDC Paris promises.

Sources: CAT (Climate Action Tracker) is average of pledges by 2030 at http://climateactiontracker. org/ assets/Global/october_2015/CAT_public_data_emissions_pathways_1Oct15.xls. UNEP is average of unconditional and conditional promises (UNEP 2015). MIT is http://globalchange.mit. edu/research/ publications/other/special/2015Outlook. LSE is the average of four different outcomes (Boyd et al. 2015). Netherlands EPA is 2015. UNFCCC is UNEP 2015. Lomborg is (Lomborg 2016) optimistic reductions using the mean of EMF27 as baseline (Kriegler et al. 2014). Climate Interactive is 2015.

level across the century. All three have about the same estimate for 2030 and see the policies in 2030 continued approximately in the same way. CAT finds much higher reductions because it assumes much larger reductions after 2030.

As such, in Figure 15.5, we see the total reduction diverge radically. MIT and Lomborg both have realistic cumulative baselines and realistic cumulative emissions with the INDCs. That is why they both find about 500–600 Gt cumulative reductions across the 21st century, which translates into about a 0.2 °C temperature reduction for both of them.

Figure 15.3 Global cumulative emissions, 2010-2100 without climate policies

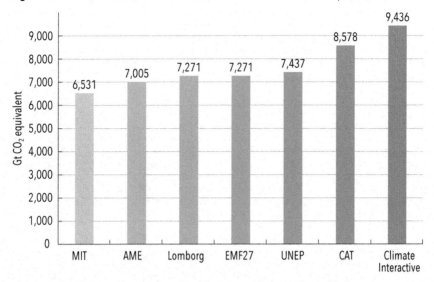

Sources: MIT is estimated based on changed CO_2 emissions from 2014–15 (MIT 2015, figure 14). AME is from Calvin et al. 2012. Lomborg is mean baseline of EMF27 (Kriegler et al. 2014). UNEP is based on UNFCCC 2015. CAT is CAT 2015. Climate Interactive is 2015.

Figure 15.4 Cumulative emissions 2010-2100 with Paris INDCs

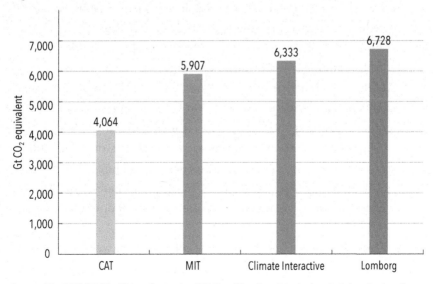

Sources: The MIT (2015), Climate Interactive (2015) and Lomborg based on optimistic reductions from EMF27 mean baseline (Kriegler et al. 2014). The CAT value includes further reductions after 2030.

Figure 15.5 Cumulative emissions reductions 2010–2100

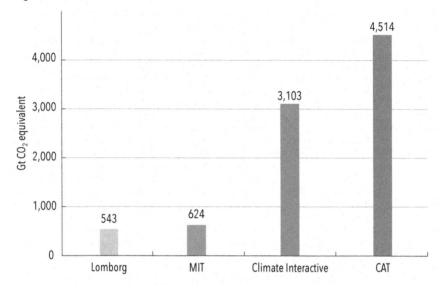

Sources: The optimistic estimates for Lomborg (2016), and also estimates from MIT (2015), Climate Interactive (2015) and CAT (2015).

CI sees about 2500 Gt higher emission reductions, but *almost all* of this reduction stems from the vastly inflated baseline. Therefore, the correct finding of CI, without the exaggerated baseline, would be almost similar to the findings of MIT and Lomborg.

CAT finds much higher reductions, yet, partly this is because it has an unrealistically high baseline (about 1,300 Gt too high, see Figure 15.3) *and* because it assumes another 2,000+ Gt reductions after 2030.

It is instructive to see how far away the estimates of CI are, since they are often used in the public discourse. In Figure 15.6, we can see the difference between all EMF24 baseline estimates, with the CI baseline increasingly diverging to the point of being about 80% too high at the end of the century.

Similarly, Figure 15.7 shows the vast difference between the median baseline of AME and the CI estimate for China emissions, which

Figure 15.6 US Business as usual emissions, Climate Interactive and 18 peer-reviewed scenarios

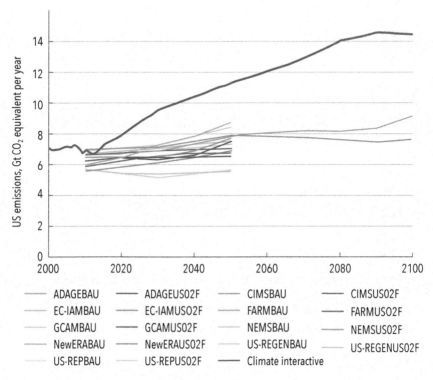

—— ADAGEBAU	—— ADAGEUS02F	—— CIMSBAU	—— CIMSUS02F
—— EC-IAMBAU	—— EC-IAMUS02F	—— FARMBAU	—— FARMUS02F
—— GCAMBAU	—— GCAMUS02F	—— NEMSBAU	—— NEMSUS02F
—— NewERABAU	—— NewERAUS02F	—— US-REGENBAU	—— US-REGENUS02F
—— US-REPBAU	—— US-REPUS02F	—— Climate interactive	

Emissions from all nine EMF24 baseline scenarios (US01 and US02) 2010–2100 (only one set goes beyond 2050), compared to the Climate Interactive estimate.

Source: Data from Kriegler et al. 2014 and Climate Interactive 2015.

increasingly diverges from the academic literature and towards the end of the century is almost 90% too high.

Conclusion

In this chapter, I have outlined both the potential temperature impact and the economic costs of the promises at the 2015 Paris Climate Summit.

Figure 15.7 China No Action, Climate Interactive versus median of peer-reviewed scenarios

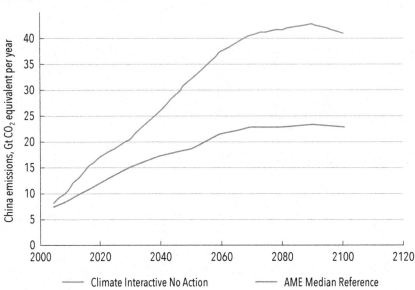

Median of baseline scenarios of seventeen individually peer-reviewed reference scenarios in the collectively peer-reviewed AME, and the Climate Interactive 'no action' scenario.

Source: Calvin et al. 2012 and Climate Interactive 2015.

What I have shown is that, by using both optimistic and pessimistic reductions in emissions, the temperature impact will do little to stabilise the climate. For instance, adopting all promises from 2016–2030 will reduce the temperature in 2100 by 0.05 °C. And further, I have demonstrated that the economic cost from pursing these objectives will cost hundreds of billions, if not trillions, of dollars in foregone economic output each year. Optimally implemented, the foregone economic output from adopting and implementing the Paris promises is estimated at US$924 billion per year by 2030. What's more, that cost will double to almost US$2 trillion per year if the policies are inefficiently implemented.

16 Re-examining Papal Energy and Climate Ethics

Paul Driessen

Scientists have determined that 'human-induced climate change is a scientific reality', the Vatican's Pontifical Academy of Sciences states in its climate change workshop report. Earth is getting warmer, fossil-fuel energy use is primarily to blame, it asserts, and unless greenhouse-gas emissions are dramatically reduced, the planet's poorest people will face 'grave existential risks' (Dasgupta et al. 2015).

The Holy Father Pope Francis's subsequent *Laudato Si, mi Signori* ('Praise be to you, my Lord') (2015) encyclical on climate change, environmental protection, and sustainable development elaborates on those claims. However, his passionate discourse is encumbered by erroneous statements, emotional contentions, and calls to action that will actually harm the human and ecological values he treasures.

Our planet 'is beginning to look more and more like an immense *pile of filth*' (ibid., emphasis added) the pontiff declares, and it simply isn't possible to sustain present levels of consumption and pollution. If current trends continue, Earth may undergo 'unprecedented destruction of ecosystems, with serious consequences' (ibid.). Therefore, 'Humanity is called to recognize the need for changes of lifestyle, production and consumption' (ibid.).

Shortly after issuing *Laudato Si*, the pontiff urged the American Congress and citizenry to join 'a conversation which includes everyone', because 'the environmental challenge ... and its human roots concern and affect us all' (Zillman 2015).

Amid increasingly rancorous energy and climate standoffs, this call for dialogue is refreshing. However, three fundamental problems cloud the Vatican's approach.

1. It accepts as unassailable fact claims that climate changes are now man-made, unprecedented, and likely to be catastrophic, requiring a complete and fundamental transformation of the world's energy and economic systems.
2. Pope Francis and his advisers have failed to participate in any dialogue with those who have different perspectives on climate change, energy, or improving the lives of Earth's poorest families. The Vatican seems more interested in lecturing people than engaging them in conversation.
3. Most important, while papal policies may protect future generations from future impacts of asserted human-caused climate and weather events, they are perpetuating serious harm to the livelihoods, living standards, and well-being of billions of people worldwide who are alive today.

The science of 'dangerous man-made climate change' is speculative

Pope Francis insists that 'carbon dioxide [CO_2] and other highly polluting gases' from fossil-fuel use are causing planetary warming, melting glaciers and polar ice caps, as well as rapidly rising sea levels, more extreme weather events and other dire problems. He believes what is happening today, or might happen over the next century, is unprecedented and that it could be catastrophic – and that humans can now control climate and weather by restricting or eliminating fossil-fuel use (Pope Francis 2015).

Many scientists support these claims. However, as other chapters herein discuss, many others disagree – raising important points that 'climate emergency' advocates cannot or will not address (see, for example, Carter 2010; Moran ed. 2015; Singer & Avery 2007; Spencer 2010; NIPCC 2014).

Climate change has been 'real' throughout Earth and human history – in response to complex, powerful, interacting *natural* forces that climatologists still do not understand sufficiently enough to guide public policy.

While CO_2 clearly does help trap solar heat and keep Earth inhabitable, it is not a 'dangerous pollutant'. CO_2 is in fact a vital plant fertiliser, essential for photosynthesis; without it plant, animal and human life on Earth would cease to exist. In conjunction with slightly warmer global temperatures, that extra CO_2 has been spurring crop, forest, grassland, and marine and freshwater plants to grow faster and better, helping to 'green' the planet (Driessen 2014a; NIPCC 2014).

A number of fundamental questions have yet to be answered:

- How do we know that the planetary warming since 1950 is primarily caused by humans?
- How accurately can we separate human influences from natural causes?
- How much will the Earth warm during the 21st century?
- At what point will further warming become 'dangerous' – and for which plant, wildlife, or human populations? Based on what convincing, empirical evidence?
- What effects are natural forces likely to have on climate over the next 50 to 100 years?
- Will slashing developed-nation fossil-fuel use actually reduce global warming by more than hundredths of a degree (which is less than the margin of error, and impossible to separate from natural shifts and fluctuations)?
- How will dramatic reductions in developed-nation emissions help, if numerous developing countries increasingly burn coal, oil, and natural gas?

These considerations cast serious doubt on Vatican claims that the 'climate crisis' is 'a moral and religious imperative for humanity' (Zaimov 2015).

These considerations also justify profound scepticism that this 'crisis' demands 'transformative solutions' that would restrict the carbon-based energy that powers the economic, transportation, health, agricultural, and communication systems that have vastly improved the 'nasty, brutish, and short' existence that once tyrannised all humanity.

Calls for dialogue are followed by dismissal of differing views

Instead of conversing with and learning from experts whose well-informed views differ from his own on energy, climate change, protecting the environment, and improving conditions for poor families, Pope Francis ignores and rejects their input.

Climate alarmists dominated the Vatican's 2015 climate workshop and ensured that scholars who question the 'dangerous man-made climate change' meme were not invited to attend, speak at, or even ask questions during the event. Indeed, there is no evidence that the Vatican ever sought or considered alternative views, or incorporated any contradictory views, before preparing the 'creation care' encyclical.

This flies in the face of the Roman Catholic Church's longstanding support for scientific inquiry.

At least since the Roman Inquisition convicted Galileo of heresy for rejecting the theory of an Earth-centred universe, the Catholic Church has been a strong defender and promoter of robust scientific inquiry.

The scientific method spread from monasteries to universities. Mankind made enormous strides in understanding natural processes, inventing new technologies, spreading knowledge and ideas, and harnessing energy and other resources to improve the human condition.

People became more aware of the essential roles we must play as responsible, accountable stewards of creation.

Echoing that proud tradition, Nobel Laureate physicist Richard Feynman famously said, 'I would rather have *questions* that can't be answered, than answers that can't be questioned' (Feynman 1985). Astronomer Carl Sagan voiced even stronger views (1980, p. 90): 'The suppression of uncomfortable ideas may be common in religion or in politics, but it is not the path to knowledge, and there's no place for it in the endeavor of science.'

Sadly, polemics, intolerance, and answers that none dare question now dominate church teachings on vitally important matters of climate change, energy development, wealth creation and redistribution, and helping the poor.

Many of the Vatican's advisers on these issues hold what many consider to be quite radical views on these topics, on population control, and on how much rich and poor nations should be 'permitted' to develop. They dominated the Vatican workshop and significantly influenced the encyclical.

The Chicago-based Heartland Institute takes a very different perspective. But when its scientists, theologians and scholars (all but one of them Roman Catholics) came to Rome for the workshop, they were forced to hold an external event and hope the Vatican would consider their presentations and an open letter to Pope Francis from the faith-based Cornwall Alliance for the Stewardship of Creation. The Vatican ignored them and made their views unwelcome. That dogmatic approach is reflected in the encyclical (see, for example, Morano 2015; Heartland Institute 2016; Cornwall Alliance 2015).

Laudato Si does call for 'academic freedom', 'forthright and honest debate', and 'discourse with all people about our common home' (Pope Francis 2015).

However, other statements suggest that Pope Francis is not really interested in discussion: 'Obstructionist attitudes' range from 'denial of the problem to indifference', the encyclical reads. The facts demonstrate that 'our common home is falling into serious disrepair'; the situation is 'reaching a breaking point'; 'the gravity of the ecological crisis' demands action on 'the spiral of self-destruction which engulfs us'; and 'we know' technologies based on using 'highly polluting fossil fuels' must be 'replaced without delay', it insists (ibid.).

Fair-minded readers might be forgiven for concluding that the Vatican and its allies want diatribes instead of dialogues, lectures instead of conversations, decrees instead of debates, and the silencing of contrarian voices within its parishes, on college campuses, and in all public forums.

That bodes ill for progress in addressing genuine environmental problems, health and income inequality, and the precarious state of our Earth's most impoverished people.

The Vatican misinterprets mankind's history of problem solving

His Holiness believes the entire world should convert to renewable energy as quickly as possible. Current policies, he says, are based on the 'false notion' that 'an infinite quantity of energy and resources are available'. Instead, mankind must implement new energy and economic systems that ensure 'sustainable development' and foster 'intergenerational solidarity' by protecting the needs of future generations (ibid.).

These views misinterpret history, natural resource use, and ethical imperatives.

For countless millennia humans endured brutal, backbreaking lives cut short by constant warfare, malnutrition and starvation, wretched cold and poverty, foul air, filthy water, a vast myriad of diseases, an absence of sanitary practices, and simple wounds that brought gangrene leading to amputation and death.

Then, in just two centuries, billions of people became healthy, well-fed, prosperous, increasingly mobile, and able to afford wondrous technologies, foods, services, luxuries and leisure-time activities that previous generations could not even imagine.

Researchers discovered the true causes of infections and diseases such as malaria, cholera, tuberculosis and polio; they discovered antibiotics, vaccinations, and pharmaceuticals to combat these infirmities and improve overall well-being, as well as anaesthesia and surgical techniques for life-saving operations. Indoor plumbing, sanitation, water purification and soap were introduced, and countless other advances were made. They learned why pollution is dangerous, and how to reduce it.

Petroleum-powered internal combustion engines replaced oxen and horses for ploughs and transportation, which in turn rid city streets of horse manure, urine and carcasses, and enabled people to travel the world and ship fresh food, clothing and other products to the globe's farthest corners in days or hours, year round.

Mechanised agriculture – coupled with modern fertilisers, hybrid and genetically modified organism (GMO) seeds, irrigation and other advances – enabled smaller numbers of farmers to produce bumper crops that feed billions, using less land, water and insecticides. Improved buildings now keep out cold, heat, and disease-carrying rodents and insects, and better survive earthquakes and extreme weather. Electricity has transformed every aspect of our lives.

The average westerner's life expectancy has soared from 46 years in 1900 to 78 today. Even welfare recipients now live healthier, longer lives than royalty enjoyed a century ago.

Over the past three decades, fossil fuels helped 1.3 billion more people get electricity and escape debilitating energy and economic poverty – more than 830 million of them because of coal. China connected 99% of its population to the grid, also mostly with coal, enabling its average citizens to be ten times richer and live 32 years longer than five decades previously (see, for example, Bryce 2014; Heerman 2015; Bezdek 2016).

This progress did not just 'happen'. It was driven by scientific inquiry, countless inventions, new corporate and free enterprise structures, improved legal, economic, property rights and communication systems, and above all coal, oil, and natural gas.

Together, those changes arguably brought greater equality and opportunity, created millions of jobs, and dramatically increased human productivity and longevity. They allowed people to specialise, innovate, and manufacture medical, agricultural and other marvels that vastly improved living standards.

Fossil fuels (and later hydroelectric and nuclear power) reliably and affordably replaced human and animal muscle, wood and animal dung, water wheels and whale oil. These fuels power factories, farms, schools, laboratories, hospitals, homes, offices, vehicles, lights and refrigerators, furnaces and air conditioners, computers, telephones and televisions, and other wondrous components of modern life.

Pope Francis himself asks (2015): 'How can we not feel gratitude and appreciation for this progress, especially in the fields of medicine, engineering and communications?'

Unfortunately, *Laudato Si* does not really appreciate the huge scope of these achievements, or how they came about. The encyclical is infused with romanticised references to bygone eras of consistent climates, benevolent natural worlds, and idyllic pastoral lives that never existed. It reprises themes that abound in Latin American 'liberation theology' dogma, in environmentalist literature, and among the Pope's Vatican and workshop advisers.[1]

---

1 Liberation theology is Christian theological movement, a blend of Marxism and Christianity, developed mainly by Latin American Roman Catholics. It argues that the plight of the poor should be central to the Christian mission and that liberation from social, political and economic oppression is the route for reducing poverty and ensuring 'social justice'.

'Never have we so hurt and mistreated our common home as in the last two hundred years,' Pope Francis laments. Cities have become 'unhealthy to live in', because of toxic emissions, noise, traffic congestion, visual pollution and general chaos. 'Once beautiful landscapes are now covered with rubbish.' People are being poisoned by industrial, agricultural and other 'toxins'. The very factors responsible for mankind's progress sometimes solve one problem 'only to create others' (ibid.).

Pope Francis continues: The scientific method is 'a technique of possession, mastery and transformation'. We should be developing renewable energy, because fossil fuels are polluting, unsustainable, and agents of dangerous climate change. Capitalist economic systems are 'focused on economic gain' and incapable of considering 'human dignity and the natural environment', whereas 'we cannot fail to praise the commitment of international agencies and civil society organizations' for the good they have done (ibid.).

The Vatican's 'solutions' would inflict far more harm than climate change

Few could deny the immensity or complexity of the human, economic, and environmental challenges facing us. However, governments and 'civil society organisations' have also helped to create or prolong miserable conditions, likely killing millions for every person who ever died at the hands of business interests. Moreover, many of the Pope's prescriptions will actually make matters worse for environmental values, poor families in developed nations, and especially the world's most powerless, impoverished, malnourished, disease-ravaged, energy-deprived people.

In 1815, one billion people inhabited our planet; today there are seven billion. Housing, feeding, clothing, transporting, and caring for all these people clearly demands more land, energy and resources, and results in more trash and pollution. It compels us to redouble efforts to solve problems and improve lives.

More than 1.2 billion people (equivalent to the population of the United States, Canada, Mexico and Europe combined) still do not have electricity. Worldwide, another two billion people have electricity only sporadically and unpredictably. In India alone more people than live in the US still lack electricity. In sub-Saharan Africa, 730 million (equal to the population of Europe) still cook and heat with wood, charcoal, and animal dung.

Hundreds of millions get horribly sick, and five million die, every year from lung and intestinal diseases, due to breathing smoke from open fires and not having refrigeration, clean water and safe food. Hundreds of millions are starving or malnourished. Millions go blind or die from dietary vitamin A deficiency. Nearly three billion survive on a few dollars a day.

With the miraculous advances of the last two centuries to guide developed and developing nations alike, this is inexplicable and unnecessary. With the uncertainties surrounding climate science, and the modern emission controls on coal- and natural gas-fired electrical generating plants, it is grating and intolerable that anyone should tell poor countries they should support their economic growth with wind and solar power – and will not receive US, European or World Bank financial aid to build coal-fired power plants. And yet, in recent years, that has happened on multiple occasions (Driessen 2014b).

Opposition to hydroelectric projects is 'a crime against humanity', a man in Gujarat, India, angrily told a UK television news crew. 'We don't want to be encased like a museum,' a Gujarati woman told the crew, in 'traditional' lifestyles so romanticised by Hollywood and radical Greens (Durkin 1997).

For non-government organisations (NGO) to tell destitute people 'that they must never aspire to living standards much better than they have now because it wouldn't be "sustainable" is just one example of the hypocrisy we have had thrust in our faces, in an era when we can and should grow fast enough to become fully developed in a single generation,' says South African anti-poverty activist Leon Louw (2002). 'We're fed up with it.'

'Cute, indigenous customs aren't so charming when they make up one's day-to-day existence,' Kenya's Akinyi June Arunga observed, following a development conference in Cancun. 'Then they mean indigenous poverty, indigenous malnutrition, indigenous disease and childhood death. I don't wish this on my worst enemy, and I wish our so-called [NGO, UN and EU] friends would stop imposing it on us' (Driessen 2005).

Developing countries know that fossil fuels are the only energy source available today to help them end poverty and disease, and to modernise and industrialise. Many supported the 2015 Paris climate treaty, or even ratified it, for two reasons:

1. They are not required to reduce their oil, natural gas and coal use, economic development, or greenhouse-gas emissions, but may do so on a voluntary basis, when it will not impair efforts to improve their people's living standards.

2. They want the free technology transfers and trillions of dollars in climate 'adaptation, mitigation and reparation' funds that now-wealthy nations promised to pay as part of what Pope Francis mistakenly calls their 'ecological debt' for damaging Earth's climate and using 'disproportionate' amounts of natural resources over the centuries.

Not only are these concerns and recommendations misplaced. Developed nations will hardly be able to pay billions of dollars annually for alleged climate adaptation and reparation if they are compelled to slash their fossil-fuel use, economic growth, and job creation.

A just and appropriate path forward

Wind and solar power can bring partial, temporary, near-term relief to remote villages, enabling people to charge cell phones, pump water, power a few light bulbs and operate tiny refrigerators. Little solar ovens and 'climate-friendly' cook stoves serve similar roles.

However, that is a far cry from the abundant, reliable, affordable electricity required for modern manufacturing, healthcare and service sectors, job creation and living standards that will help the world's poor take their rightful places among Earth's healthy and prosperous people.

Poor nations must be connected to fully fledged electrical grids, powered by coal, natural gas, nuclear or hydroelectric facilities, as soon as possible. Otherwise, people will be forced to 'enjoy' only minimal, expensive, constantly interrupted electricity, when it is available – rather than the adequate, affordable, dependable power required for modern lives, livelihoods and living standards.

Resource depletion and sustainability are not workable, ethical guide-lines for 'social justice'. They are politicised, infinitely malleable concepts used in conjunction with 'dangerous man-made climate change' to justify demands that first-world living standards be rolled back (at least for those not among the ruling elites) and that third-world lives be improved only marginally.

No one could possibly have predicted the incredible technological changes that occurred over the last two centuries, two decades or even two years, the energy and raw materials required to make them a reality, or the resource requirements they displaced.

No one predicted the horizontal drilling and hydraulic fracturing ('fracking') revolution that demolished 'peak oil and gas' worries, by providing at least a century of new hydrocarbon supplies and enabling us to extend the lives of conventional oil and natural gas fields – using a tiny fraction of the land and water that is required to grow biofuel crops.

And yet, under sustainability precepts, humanity is supposed to predict *future* technologies – and ensure that today's resource needs will not compromise the completely unpredictable energy and raw material requirements that those future unpredictable technologies will introduce.

We are likewise supposed to safeguard the assumed needs of *future* generations, even if it means *ignoring or compromising* the needs of *current* generations – including the needs, aspirations, health and welfare of the world's poorest people.

'Resource depletion' claims routinely fail to account for new technologies that increase energy and mineral production, thereby reducing their costs – or decreasing the need for certain commodities, such as fibre-optic cables, which are made from one of Earth's most abundant raw materials (silica), and which have reduced the need for copper.

Under 'precautionary', climate change and other 'principles', humanity is told it must focus on the risks of *using* fossil fuels, chemicals and other technologies – but never on the risks of *not* using them. We are required to emphasise hypothetical or exaggerated risks that a technology might cause, but ignore the risks it would reduce or prevent.

Nevertheless, those unworkable, agenda-driving principles were used to justify directives that US government agencies not help finance a US$185 million electrical plant that now provides electricity for millions of Ghanaians, using natural gas that was being burned off at the country's oil production facilities. They were employed to justify NGO and European Union demands that the World Bank not provide loans for South Africa's state-of-the-art Medupi coal-fired power plant, which is now improving millions of lives and reducing dangerous pollutants to 90% below what the 1970s-era power plants emitted (Helman 2010; Friedman 2010).

Generating just 20% of US electricity with wind power would require some 186,000 turbines, 30,577 km of new transmission lines, 18 million acres of land, and 245 million tons of concrete, steel, copper, fiberglass and rare earths – plus fossil-fuel back-up generators (Iberdrola 2008). It would also result in the deaths of millions of raptors, other birds, and bats, every year (Bryce 2015; Wiegand 2015).

America is already ploughing more than 14 million hectares of land and using billions of litres of water to grow maize for ethanol, to replace

10% of its gasoline. Converting these crops into fuel also drives up the cost of food and food aid.

Providing 50–100% wind, solar and biofuel energy worldwide would likely result in replacing land currently used for food crops, as well as devastating wildlife habitats. The wind and sun may be free, perpetual, ecologically friendly and sustainable, but the land, water, fuels and raw materials required to harness this unpredictable energy certainly are not.

Pope Francis says biotech companies employ 'risky' and 'indiscriminate' genetic 'manipulation' to advance their 'economic interests' (Pope Francis 2015). In reality, biotechnology can improve poor nations' nutrition and living standards and reduce pesticide use. It requires less land and helps countries adapt to climate change, by providing maize, wheat, cotton, and other crops that can withstand droughts.

Genetically engineered Golden Rice and bananas are rich in beta-carotene, which humans can convert to vitamin A in order to prevent childhood blindness and save millions of lives. Genetically modified (GMO) crops could help feed billions of people, without having to plough up millions more acres of wildlife habitat.

These technologies also benefit farmers. 'With the profits I get from the new Bt maize,' former South African subsistence farmer Elizabeth Ajele told the Congress of Racial Equality, 'I can grow onions, spinach and tomatoes, and sell them for extra money to buy fertilizer' (Boynes 2004).

The real moral and ethical issues

Energy, climate, and sustainability truly are critical moral and ethical issues – but not for reasons presented in *Laudato Si*. The developed world does owe an economic and ecological debt to poor nations – but not for using natural resources or causing climate change. It owes a debt and an apology for unethically employing climate and sustainability arguments to justify keeping poor nations mired in poverty, misery, disease, malnutrition and early death.

Countries worldwide are blessed with abundant oil, gas, coal and other natural resources. Turning food into fuel squanders fossil-fuel wealth and diverts land, water, fertilisers, and energy away from feeding people, to producing expensive, unreliable fuels – leaving millions malnourished. Trying to ensure future generations of materials they may not need, because of unpredictable technological advances, may deny current generations the energy and minerals they urgently need today.

Imposing fossil-fuel restrictions and renewable energy mandates – in the name of stabilising planetary climate that has never been stable – would reduce job creation and living standards in developed nations, especially for poor, minority and working-class families. Implementing policies to protect the world's still impoverished and energy-deprived masses from hypothetical man-made climate dangers decades from now would perpetuate their squalor and kill millions more tomorrow.

All these policies are short-sighted and immoral. They are unconscionable crimes against humanity – acts of callous, genocidal, eco-imperialism.

No one has a right to tell the world's poor they cannot use fossil fuels to improve their lives, or tell more fortunate families that they must reduce their living standards – based on unverified computer models and unsubstantiated fears of man-made climate cataclysms.

Prosperity and modern technology help nations better understand Earth's climate and predict future changes and extreme weather events. They enable people to adapt to warmer or colder, wetter or drier conditions; give them sufficient warning and transportation to escape looming disasters; and ensure that all people in all nations can benefit from the abundant, reliable, affordable energy that makes jobs, health, food, modern homes, better living standards, and steadily improving environmental quality all possible.

Simply put, developing nations should not do what wealthy countries are doing now that they *are* rich. They should do what those countries did to *become* rich.

Meanwhile, Pope Francis should engage contrarian voices in spirited discussion and debate; read less of Paul Ehrlich, Al Gore and other doomsayers; and delve deeply into the works of Dennis Avery, Indur Goklany, Bjørn Lomborg, Julian Simon, Matt Ridley, and their fellow realistic, rational optimists.

Above all, Pope Francis should encourage developing countries to use fossil fuels and every other source of reliable, affordable energy to carry their citizens into the ranks of the world's healthy and prosperous people, as God and his Judeo-Christian teachings desire. That would put the Pope in closer alignment with changes occurring worldwide. For example, President Donald Trump's administration will likely:

- embrace greater fossil-fuel production for the United States and developing nations; de-emphasise renewable energy and dangerous man-made climate change
- re-examine Obama-era climate data, conclusions, and policies
- insist on more robust debate, on expanded attention to empirical data and observations, and on listening to scientists who are sceptical of climate-chaos claims and the role of natural forces in climate change.

As Australia, Britain, Germany, Japan, Poland, South Korea and the US join China, India, Indonesia and other emerging economies in focusing more on coal, oil and natural gas to generate energy, jobs, health, and economic growth, that realignment would seem a wiser course than that prescribed by Pope Francis in his encyclical.

17 Free Speech and Climate Change

Simon Breheny

In these early days of the Trump administration, the new United States government has signalled its intention to change direction on climate change policy. President Donald Trump nominated for his cabinet a number of people known to be sceptical, in one way or another, about anthropogenic climate change; and the administration has already approved the construction of two oil pipelines blocked by the previous Obama administration (Sidahmed 2016). Foran, in *The Atlantic*, has complained that Trump's election was 'a triumph for climate denial' (2016).

The new administration's actions led to calls for a 'scientists' march' on Washington, DC, to protest the new approach (Worland 2017). As of 26 January 2017, more than 115,000 people had joined a Facebook group organising the protest. At the time of writing, the march was scheduled to take place in April 2017 (Ridley 2017).

Climate change public policy should be informed by a sober assessment of the data and a contest of ideas and findings, and ultimately determined by the people's representatives in consideration of the vast number of goods, both public and private, affected. But there is now an ongoing campaign by the political class, entrenched academia, and professional activists to silence scientific debate about climate change.

The idea of 'scientists, science enthusiasts, and concerned citizens' marching on the American capital can be understood in this context. It is an expression of political power in service of a mythical consensus, the assertion of which threatens the scientific method and, too often, the fundamental right to freedom of speech.

This chapter will outline the many ways in which proponents of interventionist climate change policy are attempting to foreclose debate. In resorting to these means of silencing others, they reveal their lack of confidence in their own position and their contempt for open public discourse.

State harassment of dissenters

At its highest level, the campaign against climate change free speech is nothing less than an abuse of state power by government officials. In the United States, legal officers and other elected officials have sought to use the authority of their offices to harass and silence private citizens who disagree with their preferred climate change policies.

Threatening the resource sector with legal consequences for their position on climate change is the official policy of the Democratic Party. Its party platform (2016) includes this demand:

> Democrats … respectfully request the Department of Justice to investigate allegations of corporate fraud on the part of fossil fuel companies accused of misleading shareholders and the public on the scientific reality of climate change.

This position is not merely rhetorical. In recent years, Democrats at the federal and state levels have pursued this goal by launching investigations of resource companies and the civil society organisations that support them.

In November 2015, for example, New York State Attorney General Eric Schneiderman subpoenaed ExxonMobil, seeking documents that might show the company had downplayed the risk that climate change might pose to future profits, and in so doing misled its shareholders. To justify the subpoena, in an interview with *PBS*'s Judy Woodruff (2015), Schneiderman pointed to the changes in ExxonMobil's public position on climate change between the 1980s and the present day:

> In the 1980s, they were putting out some very good studies about climate change. They were compared to Bell Labs as being at the leadership of doing good scientific work. And then they changed tactics for some reason, and their numerous statements over the last 20 years or so that question climate change, whether it's happening, that claim that there is no competent model for climate change.
>
> So we're very interested in seeing what science Exxon has been using for its own purposes, because they're tremendously active in offshore oil drilling in the Arctic, for example, where global warming is happening at a much more rapid rate than in more temperate zones. Were they using the best science and the most competent models for their own purposes, but then telling the public, the regulators and shareholders that no competent models existed?

Leaving aside the question of whether investing in profitable operations can really be said to be contrary to the interests of shareholders, this subpoena (and intention to prosecute) indicates that the State of New York now considers climate change scepticism to be dishonest.

The position of the left is now that it is not possible to disagree with their climate change position in good faith. ExxonMobil's current position of continuing to invest in fossil fuels must be at odds with its research, and therefore it is knowingly acting against the interests of shareholders.

Put another way, since anthropogenic climate change and its catastrophic consequences are facts, and they are known by ExxonMobil, the company cannot honestly take a position other than that prescribed by the Democrats and the environmental lobby. The state is saying that because it does not believe that ExxonMobil's current position is honestly held, or could be honestly held, it should not be able to hold it.

This bullying certitude is extended to groups that support the position of resource companies. As evidence for his accusation of bad faith against those companies, Schneiderman extended his criticism to a number of free-market think tanks (Woodruff 2015):

> [T]hey have made numerous statements, both Exxon officials and in Exxon reports, but also through these organisations they fund, like the American Enterprise Institute, ALEC, the American Legislative Exchange Council, through their activities with the American Petroleum institute, so directly and through other organizations, Exxon has said a lot of things that conflict with the statement that they have always been forthcoming about the realities of climate change.

The crime, then, of ExxonMobil's directors is not just believing the wrong thing, or pretending to, but giving voice to this and similar views.

And this reveals the real reason behind this investigation. It is not that the state has any concern for the well-being of shareholders – who, after all, continue to see good returns on their investments. It is that a company the size of ExxonMobil has the power to influence the public debate in a way that conflicts with the state's own priorities. It is an attempt to use the criminal law to limit the scope of the public debate about climate change, and more specifically, climate change policy (since there are a number of ways that governments might respond to even the most serious assessments of the risks of climate change).

This point was emphasised again in April 2016 when the Virgin Islands Attorney General Claude E Walker subpoenaed the Competitive Enterprise Institute requesting access to internal communications and private donor information ('CEI Fights Subpoena to Silence Debate on Climate Change' 2016).

Around that time, the attorneys general of fifteen states, the District of Columbia, and the Virgin Islands entered into a 'Climate Change Coalition Common Interest Agreement'. The terms of the pact contained a kind of *omerta*, stating that signatories should maintain confidentiality of their discussion and 'refuse to disclose any shared information unless required by law'.

The goal of the agreement was to coordinate legal actions to 'defend' federal emissions targets, and investigate 'possible illegal conduct to limit or delay the implementation and deployment of renewable energy technology' and 'representations made by companies to investors, consumers and the public regarding fossil fuels, renewable energy and climate change' (Wade 2016).

At the federal level, Democrats have been just as active.

In February 2015, US Senators Ed Markey of Massachusetts, Barbara Boxer of California, and Sheldon Whitehouse of Rhode Island sent 100 letters to 'fossil fuel companies' and 'climate denial organisations' regarding their funding of science research (see example letter to Gerard 2015).

The letters demanded information over the last ten years of all 'funded research efforts' and information about the purpose of the funding and details of the recipients.

The message is clear: how private organisations spend their resources, and how they express and develop their ideas, are now matters for the state. Cato Institute president John Allison rightly called this action 'an obvious attempt to chill research into funding of public policy projects [that the senators] don't like' (Rivkin & Grossman 2016).

This message was reinforced in July 2016, when, outrageously, Senate Democrats sent letters to 22 national and state-level think tanks demanding they disclose the identities of their donors (Mooney 2016):

> Freedom of speech does not prevent us from speaking out when your organizations, as well-funded agents of hidden principals with massive conflict of

interest, subject our constituents to an organized campaign to deceive and mislead them about the scientific consensus surrounding climate change.

As attempts to legislate against climate change have stalled, the American 'left' determined that it would no longer engage in a free exchange of ideas about climate change. Instead, it set about silencing its opponents through intimidation.

Nor is this determination to intimidate dissenters a purely American phenomenon. The Institute of Public Affairs (IPA) has also experienced the overreach of the entrenched political class. On 27 September 2006, the then Member for Wills, Kelvin Thompson of the Australian Labor Party sent a letter (2006) to a 'substantive number' of companies demanding to know whether they had given financial support to the IPA or 'any other body which spreads misinformation or undermines the scientific consensus concerning global warming', and further demanding that if they have given that support that they cease to do so.

A member of parliament has no more right to know this information, much less to make this demand, than any other citizen. And yet the demand, outrageous on its face, came in the guise of authority. This is an abuse of office and arguably a misrepresentation of the scope of Thompson's authority intended to confuse the recipient. It should also be noted that his demand extended beyond misinformation to anything that undermines the supposed consensus, which logically includes new evidence refuting it.

It is an extraordinary step for the state to take it upon itself to determine the conclusion of what should be a lively scientific debate with potentially diverse ramifications for public policy in any number of areas. To do so is to foreclose the search for truth and, moreover, to limit the exercise of the fundamental right to freedom of speech. Hoover Institution scholar Richard A Epstein wrote (2016) about these legal manoeuvres in *Newsweek*:

The usual way in which to hash these matters out is to have an intelligent debate on the pros and cons of each side. And a debate over these matters should receive the highest level of constitutional protection, given that it would be about finding out the truth, and using that information to guide political action.

But now state officials are actively seeking opportunities to punish anyone who departs from the preferred narrative. Even if these prosecutions never proceed, the state publicly contemplating such action is in and of itself an attempt to disrupt and derail discussion about climate change.

Reputation as shield and sword

Coordinated attempts to deploy the criminal law against the resource industry and organised climate change scepticism is just one aspect of the campaign to control the climate debate. Scientists themselves have worked to shut dissenting voices out of the discussion.

A notorious free-speech case involves the Canadian political commentator Mark Steyn and Pennsylvania State University researcher Michael Mann.

Mann became famous in 1998 for inventing the 'hockey stick' graph. The graph purports to show the trend of global average temperature over a 1000-year period; it suggests that there was no Medieval Warm Period. The Medieval Warm Period is when the Vikings settled southern Greenland (starting around AD 986), and corresponded with a period of cathedral building in England. This warm period was before the Great Famine and the Black Death, which occurred when it got colder in the fourteenth century. Mann's hockey stick literally flattened this history. Until the 'hockey stick', it was generally acknowledged that temperatures were about as warm 900 years ago, as they are now. Indeed, the first IPCC Scientific Assessment report (Houghton, Jenkins

Figure 17.1 Schematic diagram of global temperature variations during the last thousand years as shown on p. 202 of the first IPCC Scientific Assessment published in 1991

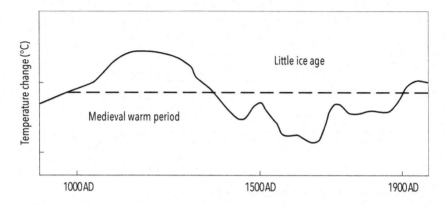

& Ephraums 1991) suggested global temperatures could have been a degree or more warmer than they are now, as shown in Figure 17.1. But by the time the Third Assessment Report was published in 2001, history had been rewritten; temperatures flat-lined until last century, when they rose suddenly resembling the shaft of a hockey stick, as shown in Figure 17.2.

In July 2012, Rand Simberg – from the American free-market think tank the Competitive Enterprise Institute – posted an article (2012) on the organisation's blog critical of Mann's employer for what he considered to be a covering up of scientific malpractice – the dubious statistics that created the hockey stick. The post deployed some very harsh personal invective in making the otherwise reasonable argument that Mann's university had not investigated claims of this alleged malpractice thoroughly enough.

A post appearing the same month at *National Review Online* by author and commentator Mark Steyn (2012) quoted Simberg's most outrageous paragraph, and, while disavowing Simberg's choice of words, supported its central contention that the investigation of Mann's wrongdoing was

Figure 17.2 The 'hockey stick' as shown in chapter 2, *Climate Change 2001: The Scientific Basis*, in the IPCC Third Assessment Report

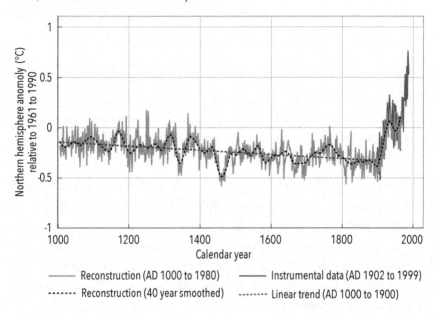

flawed. Steyn also labelled Mann's famous graph 'fraudulent'.

For reprinting Simberg's criticism, and for simultaneously disavowing the worst aspects of it, *National Review* and Mark Steyn are now being sued. In August 2012, Mann filed a defamation lawsuit with the District of Columbia Court of Appeals against *National Review* and Steyn, as well as the Competitive Enterprise Institute and Rand Simberg, for the two posts. In December 2016, after years of legal wrangling, the Washington DC Appeals Court ruled that the case may continue to trial as there was a reasonable likelihood of it succeeding.

Writing in the *Washington Post* (2016) Jonathan Adler argues that this ruling is 'unfortunate' and 'troubling'. The court held that Simberg and Steyn, based on the evidence, 'in fact entertained serious doubts' that their claims about Mann were false. This finding was based on the university's own investigation into the allegations of misconduct by Mann, which cleared him. But, as Adler notes, the adequacy or otherwise of this investigation is precisely what is at issue in the blog posts.

> It cannot be that once some official body has conducted an investigation of an individual's conduct, that further criticism of that individual, including criticism that expressly questions the thoroughness or accuracy of the investigatory body, is off limits. By this standard it would be defamatory to express the opinion that George Zimmerman or Darren Wilson is a murderer ...[1]

The threat to freedom of speech is obvious. The Mann case has established that official scientific bodies can quash dissent by effectively declaring research unimpeachable.

But the lesson to take from this is not just that the courts and academia are willing to stretch defamation law to breaking point. Nor is it that well-funded climate scientists are able to tie up opponents in years of litigation at immense cost.

It is the broader point that the tenured academic elite no longer sees any need to debate the merits of climate change science or policy and has therefore decided to shut down the conversation by any means available to them. Academia, as much as the political class, is driving the push to punish diverse views. This is a radical departure from the traditional standards of open and free inquiry.

Academic reputation is now being used not just as a shield but as a sword. It is a weapon for fighting off dissent; any sufficiently esteemed scientist is now practically beyond questioning.

---

1   Zimmerman was the man who was acquitted of the murder of Trayvon Martin; Wilson is the police officer acquitted of the murder of Michael Brown, the shooting which sparked the Ferguson, Missouri riots and subsequent Black Lives Matter protests.

Institutionalised activism

In the context of the Mann case, it shouldn't be surprising that it is academia that is leading the charge for extending the criminal law to the punishment of climate heresy.

Indeed, scientists have supported the US government bringing anti-corruption charges against resource companies. In September 2015, twenty scientists wrote to former president Barack Obama demanding that his administration use the law to go after 'deniers' (Laden 2015). And writing in the *Providence Journal*, Michael E Kraft of the University of Wisconsin–Green Bay argued the same, saying, 'Those who intentionally misled the public about climate change should be held accountable' (2016). A director at the Union of Concerned Scientists opined in the *Sacramento Bee* (Alvord 2016) that 'major fossil fuel companies' know they are contributing to global warming, so they are engaged in 'fraud' not free speech.

The creative use of anti-corruption law is not the only means by which militant climate alarmists have sought to win the debate by judicial fiat. An academic has argued seriously that resource companies could be liable for criminal negligence as a result of their 'organised campaign funding misinformation'. Philosophy associate professor Lawrence Torcello writes generously (2014) that he does not 'believe poor scientific communication should be criminalised', but nonetheless 'the charge of criminal and moral negligence ought to extend to all activities of the climate deniers who receive funding as part of a sustained campaign to undermine the public's understanding of scientific consensus'.

It is obvious that Torcello has no understanding of how negligence law actually works. Proving a causal connection between any statement or action of resource companies and a specific harm caused would be impossible, even assuming that all claims made by climate alarmists are true. Disturbingly, however, he did consider, and dismiss, the free-speech implications of his argument.

My argument probably raises an understandable, if misguided, concern regarding free speech. We must make the critical distinction between the protected voicing of one's unpopular beliefs, and the funding of a strategically organised campaign to undermine the public's ability to develop and voice informed opinions.

Just as with the harassment of ExxonMobil and the Mann defamation case, this argument assumes precisely what is in contention: the veracity of climate change apocalypticism. Despite the protestations of the accused, their conduct is assumed to not just be false, but insincere.

In a similar vein, a Norwegian academic, Lavik, has argued (2016) that 'climate denialism' should not be protected speech because it is obviously not meant sincerely and therefore does not contribute to the search for truth. This is the frightening argument – that liberalism itself does not even tolerate, let alone protect, debate about climate change.

Lavik argues (ibid.) that the utilitarian justification for free speech first propounded by John Stuart Mill does not apply to 'a well-organised and well-funded campaign by a person or group with authority in society, which keeps repeating the same untrue and damaging claims about climate change, without mentioning conclusive counter-arguments'. This is so because while freedom of speech generally advances the pursuit of truth, this is not true for utterances that are not sincerely meant.

Not stated is how it is that the truth of a claim, and the sincerity behind it, are to be judged. By what principle might a threshold of sincerity be set, above which speech is permitted and below which it is prohibited? The establishment of that standard is itself to make a truth claim and therefore to prevent debate about whatever is in contention.

In short, granting the state plenary power to determine the truth is to unwind the Enlightenment. No dissent from established truths could ever be tolerated.

For this reason, it is not an exaggeration to say that climate change alarmists are attempting to establish themselves as a kind of clerisy. If that sounds hyperbolic, consider the following argument, presumably made sincerely, by a pair of Australian academics (Shearman and Smith cited in Berg 2007):

> [T]here is some merit in the idea of a ruling elite class of philosopher kings. These are people of high intellect and moral virtue ... These new philosopher kings or ecoelites will be as committed to the value of life as the economic globalists are to the values of money and greed.

An elite class of people to protect unimpeachable truths, to proselytise the good word to the unconverted, and to tend to the needs of the initiated – climate change has taken on a quasi-religious quality. It shouldn't surprise then that since climate change sceptics are dissenters from the high church of scientism, they must be silenced and banished.

All hail the philosopher kings

The elite cabal of scientists who dominate climate change research have been militant in shunning those with whom they disagree. Their instinct is to protect their turf rather than to engage in free debate. This contributes to a powerful chilling effect that, in concert with the state's campaign of harassment, discourages dissenters from voicing their opinions.

For example, the platforms that are otherwise given to academics and experts are denied to dissenters, no matter their credentials. Danish statistician Bjørn Lomborg came to prominence in 2001 with his book *The Sceptical Environmentalist*, in which he argued that many policies advocated by climate change alarmists would be ineffective and wasteful. In 2015, Lomborg agreed to move his research centre to the University of Western Australia, with funding coming from the Australian government. But under pressure from academic staff and student activists, the university's vice chancellor pulled out of the deal. Subsequent discussions with Flinders University floundered immediately as staff and students reacted angrily to the idea that someone might ever produce research they disagreed with (Hasham 2015).

In Portland, Oregon, the public school board voted to ban any textbooks that expressed any doubt at all about the causes or severity of climate change. Free-speech advocates responded by making Mill's point that even if you agree with climate change alarmism, you should want people to be familiar with the contrary arguments if only to strengthen your own (Flood 2016).

But as we have seen, the utility of freedom of speech is now doubted by the left, especially as it relates to climate change. The academic cartel is backed in its attempts to chill debate by a pliant media class that seems to have forgotten the role said freedom plays in their livelihoods. The argument they make is that whatever the value of free speech might be, climate change is sufficiently grave a threat to outweigh it. One Australian commentator, Judith Brett, has written (2014) that:

> [F]reedom of speech is indeed a cornerstone of our democratic society. But in the current politics of climate-change action, to advocate this kind of view so strenuously contributes to the weakening of the political will needed to mitigate the risk of climate change.

The author goes on to make a somewhat bizarre analogy. Speaking about Australian Attorney-General George Brandis, who had commented that there was value in allowing dissenting voices to be heard, she writes:

> Brandis' argument that we should always consider the possibility of an alternative opinion is epistemologically disingenuous. I doubt that Brandis believes that all alternative points of view are deserving of respectful consideration. I doubt that he believes that the Earth is flat or that carrot juice can cure cancer. I'm sure that when he boards a plane he believes that the science of aerodynamics is sufficiently settled to get him to his destination.

The argument, of course, elides the difference between permitting arguments to be made and actively endorsing them. It is true that no one boarding a plane is likely to pay much heed to alternative theories of aerodynamics, but it does not follow that those theories ought to be banned.

The silencers place great weight on the threat of climate change. The stakes of this debate are held to be too high for all viewpoints to be aired. But, again, this is to beg the question. The threat of climate change is what is in dispute. Declaring any piece of evidence, no matter how compelling, to be dispositive is to shut down the debate. As such, what is truly disingenuous is the claim by climate change alarmists that they have any concern whatsoever for the rights of others.

This lack of respect taints the entire public discussion of climate change. Proponents of climate change action routinely use the word 'denier' to describe climate change sceptics, with the word having been deliberately chosen to invoke a parallel with Holocaust denial. As Brendan O'Neill has argued (2006):

> There is something deeply repugnant in marshalling the Holocaust in this way ... [the Holocaust] is an historical event that has been thoroughly investigated, interrogated and proven beyond reasonable doubt. [Whereas] the turning of climate change denial into a taboo raps people on the knuckles for questioning events, or alleged events, that have not even occurred yet. It is pre-emptive censorship.

The very framing of the debate is slanted. Not only do sceptics have to contend with harassment from elected officials and the scorn of their academic peers, they are slandered as soon as they open the mouths or type a single word.

Universities are taking steps to perpetuate this bias in scientific commentary. Students can now undertake a bachelor degree in science communication. These students can look forward to a life churning out listicles of rehashed talking points and propagandising the nostrums of the oracles. Rather than being trained with the scientific or critical reasoning skills they would need to participate in debate, they are being trained only to parrot received wisdom.

If climate change scientists are the new philosopher kings, then they have already established a substantial court. Their fellow academics have argued in defence of their exalted status and the media has shaped their public image so as to entrench their influence. Together, the courtiers represent a malign influence on our culture of free speech; the power they have to chill and deter debate through intimidation and misrepresentation threatens not only the rights of dissenters but the pursuit of truth itself.

Conclusion

Freedom of speech is a basic human right. Individual autonomy is meaningless without the right to express how one thinks and feels. It is a right that we all owe to one another, based on nothing more or less than our mutual recognition as beings of inherent value.

But freedom of speech is not only valuable in and of itself, it has instrumental value as well. It is only through the contest of ideas that the truth may be apprehended. In science, as in politics and all other fields of human endeavour, dissent has the virtue of testing the prevailing arguments, which either fail and fall, or survive yet stronger. The peremptory shutting down of debate short-circuits this process.

It might be thought that scientists would grasp this argument intuitively. After all, their discipline is the lodestar of empiricism, and the truth that it produces depends on the testing and re-testing of evidence and data continuously accumulated.

The issue of climate change is fundamentally one of science, yet if there were any doubt that climate change has moved beyond the dispassionate search for truth through empirical enquiry and into the realm of political activism, it has by this point surely been exhausted. And as the issue has mutated from science to politics, dissenting opinion is tolerated less and less. From state attempts to use the criminal law against company directors, to demands for the use of anti-corruption law to criminalise scepticism, to invidious defamation actions to chill it – proponents of climate change have undertaken to silence their opponents by any means necessary. This intolerance also includes the activism of the bureaucracy and academia, and, as the Trump administration will doubtless discover, the combination of these factors represents a concerted attempt to close off the possibility of alternative views prevailing.

The debate over climate change has taken a sinister turn, away from the pursuit of truth and towards the propagation of dogma. Climate change scientists have encouraged the state to interfere in their work and they have contributed to an assertion of academic privilege that is more befitting of a religious hierarchy than rationalism.

Should scientists and their bannermen march on Washington DC, they will do so in order to express their opposition to the beliefs and choices of the new Trump administration. They will be exercising their right to dissent from those in positions of power, from those with the ability to shape the public debate about climate change and the policy response to it. They will be claiming for themselves the right to promote their sincere concerns and their right to organise in defence of those concerns.

And as they march, it is to be hoped that the irony is not lost on them.

18 The Lukewarm Paradigm and Funding of Science

Dr Patrick J Michaels

Increasingly, our world is one of global 'lukewarming', where global warming is real, but neither momentous now, nor likely to become so, within a foreseeable future. The surface average temperature of our planet is about 0.8 °C warmer than it was at the turn of the twentieth century. Human beings have something to do with this, but they most certainly don't have *everything* to do with it. It is also important to note that, if United States temperatures are any guide, the pace of recent warming is likely to be overestimated (Watts et al. 2015). Other, truly independent, sources – microwave-sounding satellites and thermistors attached to ascending weather balloons – confirm that warming is real, and establish that it is occurring at only a modest pace. A modest warming, which is only partially fuelled by human activities, during an era of rising atmospheric greenhouse-gas concentrations, serves to elevate support for lukewarming, and to downgrade worries of rapid and alarming, temperature increases.

Modelled climate sensitivities in the latest suite of the Intergovernmental Panel on Climate Change (IPCC) general circulation models (GCMs) (Stocker et al. 2013) clearly average too high. The climate community, which is defined by models and modellers, is engaged in the type of behaviour predicted by Thomas Kuhn in his classic book

The Structure of Scientific Revolutions (1996). This community generates data in support of a paradigm that may violate basic physics, blatantly cherrypicking to support the policy science. The incentive structure in modern science requires that practitioners largely support the high-sensitivity model-based paradigm of warming in order to remain employed and to advance in their careers. The result is a polluted canon of knowledge. This disorder is systemic across most areas of science that are difficult to replicate, climate science being a prime example.

Ignoring anomalies in defence of the existing paradigm

A key uncertainty in projecting future climate change is the magnitude of equilibrium climate sensitivity, which is the eventual increase in global annual average surface temperature in response to a doubling of atmospheric concentration carbon dioxide (CO_2).

Around the time that the IPCC's Second Assessment Report was written, in 1995, the first generation general circulation models (GCMs), incorporating both greenhouse warming and sulphate cooling, were published. The equilibrium climate sensitivity of most of these did not differ much from the most recent suite of 30 models included in the IPCC's Fifth Assessment Report (Stocker et al. 2013), with an average sensitivity of 3.2 °C. This figure is higher than the average sensitivity of 2.0 °C found in a series of papers published from 2011, as detailed in my contribution to the previous volume of *Climate Change: The Facts* (Michaels 2015). I also published a much earlier paper (Michaels et al. 2002) implying a sensitivity of approximately 1.6 to 1.9 °C.

All of these low-sensitivity publications have been largely ignored. A particularly prescient passage in Kuhn's book (1996) describes this as the normal situation in science:

> In science ... novelty emerges only with difficulty, manifested by resistance, against a background provided by expectation. Initially, only the anticipated

and the usual are experienced even under circumstances where anomaly is later to be observed.

Besides the publications mentioned above, there are several more 'anomalies' that are being ignored, or are somehow considered to be consistent with the high-sensitivity model-based paradigm.

Figure 18.1 shows the large and growing mismatch between the global average temperature evolution in the 2013 IPCC suite of GCMs, along with what has been observed by weather balloons and satellites.

Figure 18.1 A comparison of climate-model simulations with observed temperature

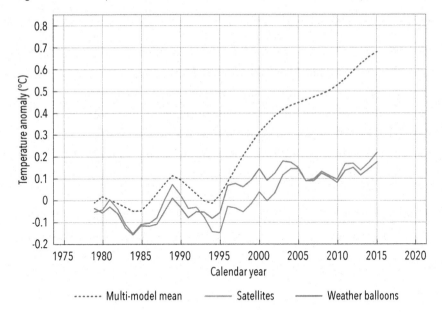

Centred five-year averages of temperatures from the middle troposphere showing the mean from 102 climate model runs (red series), three satellite data compilations (green series), and four weather-balloon datasets (blue series).

Source: adapted from Christy, JR 2016, 'Testimony to the U.S. House Committee on Science, Space & Technology', 2 February, viewed 28 March, 2017, http://docs.house.gov/meetings/SY/ SY00/20160202/104399/HHRG-114-SY00-Wstate-ChristyJ-20160202.pdf.

It is a stark testimony to the Kuhnian stickiness of the high-sensitivity model-based paradigm that these anomalies become 'the anticipated and usual'. Critics argue that the satellite temperature is not relevant because it does not reflect conditions at the Earth's surface. This is a remarkable defence of the existing paradigm, as incorrectly specifying the vertical distribution of temperature invalidates virtually every model-based simulation of cloudiness, convection, and precipitation, that are indeed all driven by the vertical temperature lapse rate.

To what lengths will scientists behave in a misleading fashion to maintain the high-sensitivity paradigm? Will paradigm-defining scientists cherrypick data to indicate everything is 'anticipated' and 'usual'?

Cherrypicking beginning and end dates

In 1996, the Fourth Conference of the Parties to the United Nations' 1992 Framework Convention on Climate Change was scheduled to begin in mid-July. Earlier that month, a remarkable paper appeared in *Nature*, one of the highest-impact science journals in the world, claiming that GCMs were very accurate in simulating the time-evolving three-dimensional thermal structure of the atmosphere as greenhouse-gas – mainly CO_2 – concentrations increased (Santer et al. 1996). The modelled atmosphere included both CO_2 warming and sulphate cooling (as well as a small effect from stratospheric ozone depletion). The core model result is shown as Figure 18.2. The signal characteristic of it is that warming is greater in the free troposphere in the Southern Hemisphere, owing to the virtual lack of atmospheric sulphates south of the thermal equator.

The peculiarity of the paper was that it covered the period from 1963 to 1987, although the upper-air data required for a three-dimensional analysis was reliably catalogued back to 1957 – by one of the paper's thirteen authors – Abraham Oort of the Geophysical Fluid Dynamics Laboratory in Princeton. The starting date of 1963 was also a very cool

Figure 18.2 Climate-model projections of atmospheric temperature change

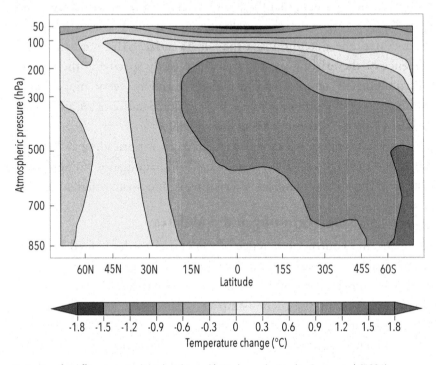

Projections for different atmospheric heights and latitudes as depicted in Santer et al. (1996).

Source: reprinted by permission from Nature Publishing Group – Santer, BD et al., 'A search for human influences on the thermal structure of the atmosphere', *Nature*, vol. 382, pp. 39-46, copyright 1996.

point in global records, as temperatures were chilled by the 1962 eruption of Indonesia's Mount Agung, one of the four large stratovolcanoes in the twentieth century, and the biggest since Alaska's Katmai in 1912.

The year 1987 also seemed to be an odd ending point. Data were certainly available through to 1994, seven years later, and updatable through to 1995. It is noteworthy that 1987 was an El Niño year, and therefore relatively warm compared to the rest of the study period.

The match between the observed three-dimensional temperature profile and the modelled profile was persuasive because of the
projected difference between warming in the two hemispheres, with a
substantial 'hot spot' – both simulated and observed – in the lower and
mid-tropospheric Southern Hemisphere, as shown in the upper chart
of Figure 18.3. And, indeed, as shown in the lower chart of Figure 18.3
(circled), the relative rise in temperature in this region during the study
period (1963–1987) was profound.

However, the omission of data from the years 1957–62 and 1988–95
was puzzling. The reason these data were not included became clear
when I added them in (Michaels & Knappenberger 1996). If all the
data were used, there would have been no significant match between
the modelled and observed data (Figure 18.3, lower chart). Santer et al.
simply discarded the data that didn't fit their preconceived hypothesis.
When our result appeared in *Nature*, Santer et al. penned a rambling and
unsubstantive response (1996).

Adjusting away the pause

Another 'anomaly' that is of particular interest to the sensitivity argument
is the 'pause', or 'hiatus', in surface warming that appeared in both surface
and tropospheric records beginning around 1996 and continuing until
the very large-scale 2015–2016 El Niño.

The temperature trends, however, were modified for the official record
by Karl and colleagues (2015). After making the changes, Karl et al. wrote,
'It is also noteworthy that the new global trends are statistically significant
and positive at the 0.10 significance level for 1998–2012 ...' But the 0.10
significance level is hardly normative science, which has long embraced the
0.05 level (a 1-in-20, or less, chance of the effect being described being due
to change (unexplained processes) as the de facto standard).

Karl et al. managed to warm temperatures during the pause by using
data that 'disinclude' (their word) satellite-sensed sea-surface temperature

Figure 18.3 Observed atmospheric temperature trends

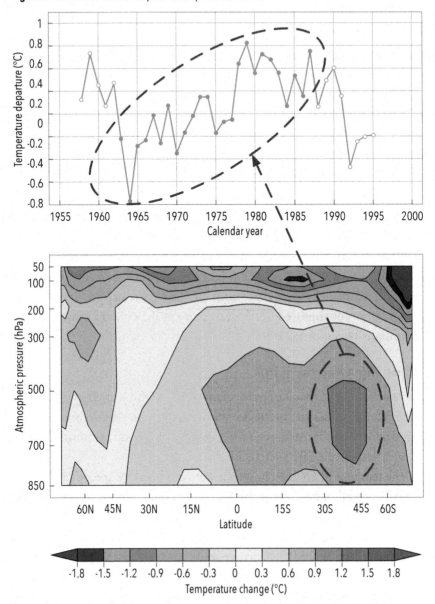

Actual trends vary with the start and end date. In the colourful area-chart (bottom), temperatures have been calculated for a component of the atmosphere for the period 1963-1987. Yet the data is available for the longer period, 1957-1995: as shown in the top time-series chart.

Source: adapted with permission from Nature Publishing Group – Michaels, PJ & Knappenberger, PC, 'Human effect on global climate?' *Nature*, vol. 384, pp. 522-523, copyright 1996.

data (2015). Instead, they used highly suspect temperatures measured in the cooling-intake tubes of ocean-going vessels. The data from a rapidly proliferating network of drifting buoys with excellent instrumentation were then adjusted upwards by 0.12 °C in order to match the ships' data. Because there are increasing numbers of buoys, this would necessarily induce some warming into recent years' data. Immediately thereafter, NASA adopted the Karl et al. (2015) modification, and similarly attenuated the pause.

A variety of concerns have been raised about the validity of the Karl et al. rewriting of the surface temperature history – concerns which led to an investigation by the US Congress (Smith 2015, pers. comm. 14 July).[1]

Stratospheric cooling

Perhaps the most intriguing issue involves the recent evolution of temperatures in the stratosphere. The theory of surface warming being created by increasing concentrations of lower atmospheric greenhouse gases requires a compensatory drop in stratospheric temperatures, as the solar energy incident upon the Earth remains nearly constant.

Figure 18.4 shows the stratospheric temperatures (after removing internal natural variability) as sensed by satellites, beginning in 1979. The obvious spikes are associated with the 1983 and 1992 tropical stratovolcanoes (El Chichon and Pinatubo), and the expected decline in temperatures is obvious *until precisely the time in which the pause in surface warming commences in 1996* – then the decline stops in this layer. The timing of the observed decline is not consistent with the ozone depletion caused by chlorofluorocarbons as that is centred higher, peaking in the mid-stratosphere. The cessation of lower stratosphere cooling coincides with the beginning of the 'pause' or 'hiatus' in lower

---

1   https://cdn.arstechnica.net/wp-content/uploads/2015/10/rep_smith_noaa_letters.pdf.

Figure 18.4 Lower stratospheric temperatures

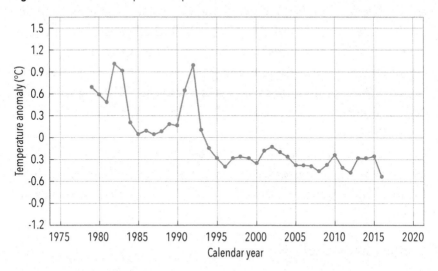

Source: John Christy, University of Alabama, Huntsville.

atmospheric warming. This is evidence that the 'pause' was a real aspect of the climate's behaviour, and not merely a result of data handling, as suggested by Karl et al. in their 2015 paper.

Modelling climate sensitivities

In the latest suite of climate models in the scientific summary of the UN's IPCC, the mean sensitivity of surface temperature to a doubling of CO_2 is 3.2 °C, and the warming rates are clearly too large. There are two independent, but mutually consistent, estimates of a revised sensitivity that roughly cut the IPCC's figure in half and bring it more in line with the observed temperature history.

The first owes to the fact that the increase in atmospheric CO_2 is a low-order exponential function, while the response of temperature to incremental changes in CO_2 is logarithmic. That this sums roughly to a straight line of constantly increasing temperature is very clear from

Figure 18.5 Global temperature projections relative to 1986–2005

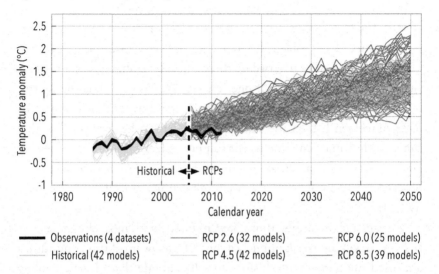

This chart shows the IPCC projections for global temperature increase with the output varying considerably depending on the mix of models used.

Source: Stocker et al. 2013.

inspection of the most recent suite of climate models (see Figure 18.5) *as well as in observed temperatures* since the second warming of the twentieth century began in 1977. The observed surface temperature history, which includes the very warm 2015 data, shows the profound linearity of global warming, which is about 60% of what has been predicted, therefore yielding a sensitivity of just more than half of the model average, or about 1.8 °C.

Sulphates do not explain lack of warming

The IPCC has offered two explanations for the lack of modelled warming, one of which is the argument concerning sulphate compensation. Sulphates occur as aerosols, largely from the combustion of coal. One of their effects in the atmosphere is to scatter incoming light, increasing the Earth's albedo, which is thought to contribute to cooling.

The IPCC explanation, however, is weak, as it contains very large uncertainties. The problem lies not with the recent warming, but with the warming of the early twentieth century. Greenhouse-gas emissions were relatively low then, so if one (erroneously) insists on ascribing this warming to human activity (which the UN explicitly did in the 2015 Paris Agreement), then the sulphate effect had to be very small at that time. At the same time, the recent sulphate effect would have to be very large in order to compensate for the extremely large warming that must be occurring now if the early twentieth-century warming were caused by CO_2.

Max-Planck Institute's Bjorn Stevens (along with most climate scientists – who will not speak publicly about it) was aware of this conundrum and used the temperature history to self-calibrate the sulphate cooling effect (Stevens 2015). In doing so, he whittled down the cooling exerted in the temperature record from sulphates to between 0.2 and 0.8 °C. Nic Lewis and Judith Curry examined Stevens's statistics and showed that the most likely value is around 0.4 °C (Lewis 2015). This is less than half of the cooling they exert, on average, in the IPCC's models.

The reluctance to abandon the modelled paradigm

Why does the climate-modelling community elide such simple and obvious calculations? The answer may be in the nature of modern science, especially modern climate science.

In the US, the academic reward structure revolves around an individual creating a substantial body of research, beginning in graduate school, and in around five years' time achieving their first 'tenure track' position as an assistant professor. In climate science, it is almost impossible to do this without a massive amount of research support, and the only provider of this is the federal government. A reasonable estimate is that it takes about US$5 million in research funds to publish sufficient research to receive tenure at a tier-one research university.

While agencies like the US National Science Foundation may claim that their awards are strictly based on peer-reviewed merit, the peer-reviewers have their own self-interest at heart, and are not likely to look kindly on proposals challenging the paradigm that fed and promoted them. Other funders, such as NASA, the US Department of Energy, and the US Department of Commerce (where Karl's revised temperature history originated), clearly have political masters, missions, and agendas. The US Global Change Research Program (USGCRP), a consortium of all federal entities that receive substantial climate change research funding, is required – by statute – to summarise the research it largely funds every five years. Who could possibly believe that it will seriously entertain grant proposals hypothesising that global temperatures are systematically flawed, that the sensitivity of surface temperature to CO_2 has been overestimated, or that people will obviously adapt to climate change (if only their economies are free enough to be sufficiently nimble)?

Consequently, the young researcher seeking tenure is forced to apply for funding within the political programme, as well as within the scientific paradigm. And who of them are going to even attempt to publish results that might imply that the USGCRP – which has brought financial security for so many – is a massive irrelevancy?

Ultimately, this leads to research results in a flux of 'publication bias' in which findings that do not support a funded hypothesis are not reported, or, if such a result is obtained, then the models and procedures are run and re-run with different or modified variables (so-called 'model tuning') until what Hourdin (2017) calls an 'anticipated acceptable range' of model output.

The evidence for this is compelling. John Ioannidis first documented such problems in biomedical science, in a paper titled 'Why

Most Published Research Findings are False' (2005) and Daniele Fanelli (2012) generalised them. Ioannidis noted that the demand to publish and get funding is so strong that many studies are designed to inevitably produce positive results, or for the experimental data to support the original research hypothesis, which was usually generated itself by a research grant proposal.

In 2012, Daniele Fanelli published a seminar paper with the title, 'Negative Results are Disappearing from Most Disciplines and Countries' (ibid.). Based on a sample of 5000 published papers, it is clear that there is a large and systematic increase in the percentage of studies reporting positive results in support of the original (and probably the research grant) hypothesis. Fanelli notes that the malaise does not appear to infect purely physical sciences, which are derived largely from mathematics, but that the problem accrues more where results are difficult to test. Given that, ultimately, climate models are about an unrealised future, it should not be surprising if the 'positive results' bias were especially operative. As an overlay, the more a country adopts the 'American model' for advancement (requiring massive funding for continued employment), the more disproportionately positive results are published. Together with Ioannidis, Fanelli found that if an international team of scientists invites an American author onboard, the probability of a positive result *doubles*.

The increasing dominance of positive results enhances the longevity of existing paradigms. Science historian David Wojick has demonstrated that the reigning paradigm in climate science is defined by climate models, even though they are clearly failing. Using the search term 'modeling', he found that 55% of *all* scientific papers employing a quantitative model are in the field of climate science, despite the fact that climate science comprises only 2% of the US science budget.

This ensures all manuscripts submitted for review that demonstrate the models are failing will receive a vigorous and largely negative review. An attack on climate models is, simply put, an attack on climate science.

Conversely, manuscripts in support of climate models, or using their output to drive subsidiary models, are likely to be lightly reviewed and quickly published.

This is in addition to publication bias. The need for paradigm protection by rejecting papers that themselves reject model hypotheses is the best way to assure that programmatic funding continues, both at the federal level and at the level of the individual researcher in pursuit of global leadership and a full professorship.

19 The Contribution of Carbon Dioxide to Global Warming

Dr John Abbot & Dr John Nicol

The appointment of Scott Pruitt by United States President, Donald Trump, as the Administrator of the United States Environmental Protection Agency (EPA) has been controversial. Mr Pruitt has been quoted as stating he does not believe that carbon dioxide (CO_2) is a primary contributor to global warming, and was reported by CNBC (DiChristopher 2017) as saying:

> I think that measuring with precision human activity on the climate is something very challenging to do and there's tremendous disagreement about the degree of impact, so no, I would not agree that it's a primary contributor to the global warming ... We need to continue the debate and continue the review and the analysis.

This statement contradicts the public position of the agency Mr Pruitt now leads. The EPA's webpage on the causes of climate change states, 'Carbon dioxide is the primary greenhouse gas that is contributing to recent climate change' (EPA n.d.). A view in accordance with the position of the Intergovernmental Panel on Climate Change (IPCC).

This chapter briefly details some of the evidence suggesting that there is considerable uncertainty in the magnitude of the contribution of CO_2 to global warming. Given the economic importance of this relationship in policy determinations, it is imperative that further analysis and investigations be undertaken. In particular, there is a need to resolve the discrepancy between measurements from modern spectroscopy (Barrett 2005; Lightfoot & Mamer 2014; Laubereau & Iglev 2013), and the approximations used in the IPCC General Circulation Models, which are based on a speculative 1896 paper by Svante Arrhenius (Arrhenius 1896).

Greenhouse gases, spectroscopy and Jack Barrett

Some of the incoming radiation from the Sun is reflected by the Earth's surface and the atmosphere. Some is absorbed by the Earth's surface, causing it to warm, as shown in Figure 19.1. Because the Earth is much colder than the Sun, it radiates at much longer wavelengths – primarily in the infrared region of the electromagnetic spectrum.

Figure 19.1 The greenhouse effect

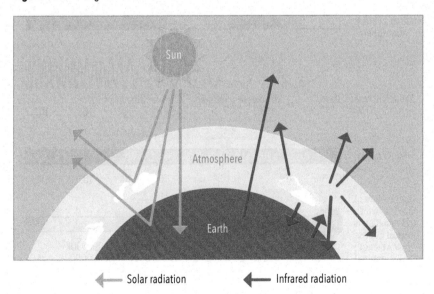

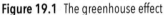

← Solar radiation   ← Infrared radiation

Some of the infrared radiation emitted from the Earth's surface passes through the Earth's atmosphere, while some is absorbed and re-emitted by greenhouse gases.

The electromagnetic spectrum is the collective term for all frequencies of electromagnetic radiation, as shown in Figure 19.2. This spectrum extends from below the low frequencies used for radio communication to include the visible wavelengths from the Sun – the infrared region is in between.

Some of the infrared radiation emitted from the Earth's surface passes through the Earth's atmosphere, while some is absorbed and re-emitted by greenhouse gases. This absorption, which warms the lower atmosphere – and the subsequent re-emission, which warms the Earth – leads to what is known as the greenhouse effect, without which the lower atmosphere would be significantly cooler.

Figure 19.2 The electromagnetic spectrum

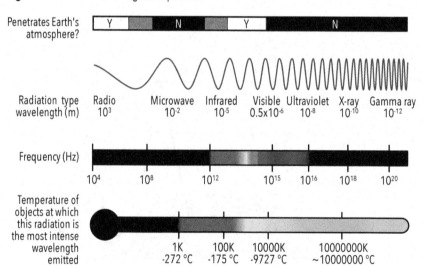

Infrared radiation is part of the electromagnetic spectrum.

Source: NASA self-made, Wikimedia Commons, viewed 30 May 2017, https://en.wikipedia.org/wiki/Electromagnetic_spectrum. Reprinted under Creative Commons CC-Share Alike 3.0 Unported license.

The hypothesis that the Earth's radiation is continuously absorbed and emitted by atmospheric CO_2, contributing to global warming, was first expounded by Joseph Fourier in 1824. Technical understanding of the precise relationship between CO_2 and the absorption and radiation of this energy has been refined ever since (Fourier 1824, 1827; Lorentz 1897; Ekholm 1901), particularly with the development of high-resolution spectroscopy equipment in the 1960s (Kuhn et al. 1967; Lewis et al. 1968; Malvern et al. 1975).

Spectroscopy is essentially the study of electromagnetic radiation, and is important in categorising its interactions with matter. High-resolution techniques for measuring the spectra of visible light were not developed until the middle of the twentieth century. Importantly, this work was extended to the infrared region of the spectrum through the seminal work of Dr Jack Barrett at Imperial College, London, in the 1980s (Barrett 1985). Barrett's work is considered particularly important for accurately quantifying the molecular absorption, radiation, and collisional interactions that define the environment of an atmospheric greenhouse gas – particularly CO_2 (Barrett 2005).

Research by Barrett arguably provided the first detailed and quantified descriptions of how CO_2 absorbs infrared radiation (Barrett 1985). This research showed that the CO_2 spectrum is dominated by bending vibrations centred at 667 cm^{-1} and anti-symmetrical stretching at 2349 cm^{-1}, modulated by a large number of different rotational energy states. Together, these provide broad absorption bands and complementary radiation bands centred on these frequencies.[1]

---

1   To clarify the spectroscopic terminology, the wavelength is the distance between the peaks of a wave, the spatial frequency is the number of wavelengths within a given unit distance and the frequency (temporal) is the number of wave peaks per second observed at a point. Thus, in the infrared spectrum the wavelengths are of the order of 1 μm to 200 μm, the spatial frequencies of order 50 cm^{-1} to 10,000 cm^{-1} while the frequencies range from 1.5×10^{12} Hz to 3×10^{14} Hz. Hertz (Hz) refers to one cycle per second.

Water vapour, methane, nitrous oxide and ozone are also considered greenhouse gases. This is because of similarities in their internal electronic structure, and the nature of the various individual atoms from which they are formed. The presence of more than two atoms in each of these molecules, allows them to vibrate and rotate, under the action of the resonant forces provided by the infrared radiation, which polarises the electric charges within the molecules.

The structure of the simpler, diatomic molecules, for example nitrogen (N_2) and oxygen (O_2), does not provide for such polarisation at these infrared wavelengths, precluding interactions with radiation emitted from the Earth's surface. N_2 and O_2 are, therefore, not greenhouse gases even though they account for more than 99.9% of the dry atmosphere.

Climate change, carbon dioxide and Svante Arrhenius

While it was Fourier in 1824 who first determined that the Earth absorbed visible light and emitted infrared radiation, the hard core of anthropogenic climate change theory is embodied in the mathematics of Swedish chemist Professor Svante Arrhenius. In a seminal paper published in 1896, Arrhenius stated that if the quantity of carbonic acid (H_2CO_3) increases in geometric progression, the augmentation of the temperature will increase nearly in arithmetic progression. According to these calculations, a doubling of atmospheric CO_2 would cause a global temperature rise of 5 to 6 °C (Arrhenius 1896).

These somewhat rudimentary calculations are *not* based on a detailed experimental understanding of the spectroscopy of CO_2, but rather on a theoretical extrapolation of work by American Professor Samuel Langley. Specifically, Langley had attempted to determine the surface temperature of the Moon by measuring the infrared radiation leaving the Moon and reaching the Earth. Arrhenius's much quoted 1896 paper – with its calculations subsequently incorporated into the IPCC's General

Circulation Models – is unashamedly an extension of this work. In particular, Arrhenius states:

> In order to get an idea of how strongly the radiation of the earth (or any other body of the temperature +15 °C) is absorbed by quantities of water-vapour or carbonic acid in the proportions in which these gases are present in our atmosphere, one should, strictly speaking, arrange experiments on the absorption of heat from a body at 15° by means of appropriate quantities of both gases. But such experiments have not been made as yet, and, as they would require very expensive apparatus beyond that at my disposal, I have not been in a position to execute them. Fortunately there are other researches by Langley in this work on 'The Temperature of the Moon,' with the aid of which it seems not impossible to determine the absorption of heat by aqueous vapour and by carbonic acid in precisely the conditions which occur in our atmosphere.

It wasn't until 1988 that this speculative theory captured significant political attention. That was when climatologist James Hansen, in his testimony to US congressional committees, claimed a 4.2 °C global temperature increase would result from a doubling of atmospheric CO_2. He was quoting output from computer models, which from the 1960s had incorporated a version of Arrhenius's original calculations.

Variations of the global temperature are typically referenced relative to values estimated in 1750; the difference $\delta T = T_{now} - T_{1750}$ being featured in the expression $\delta T = \lambda.RF$.[2] Since Professor Hansen's 1988 testimony, these estimates have been continuously adjusted down. In the 1995 IPCC report, for example, a doubling of CO_2 was predicted to cause a 3.8 °C increase; then in 2001 a 3.5 °C increase; and in 2007

---

2   Where RF, the radiative forcing, is defined by the estimated difference between the net downwards infrared radiation at the tropopause in those two different years (now and 1750), the tropopause being that part of the atmosphere above cloud level, $\lambda = 0.75$, is a constant referred to as the 'climate sensitivity parameter', and RF ≈ 3.5 at the present time.

a 3.26 °C increase. In 2008, twenty years after his initial influential testimony, Professor Hansen issued a statement to the effect that his central estimate for lambda (λ) was now 0.75, requiring a further reduction of the official climate sensitivity estimate by one-quarter, to 2.5 °C degrees for a doubling of CO_2.

The role of carbon dioxide in radiative forcing – according to the IPCC

The IPCC uses the term 'forcing' to summarise how the Earth's radiative balance has been perturbed from its natural state, including by increasing concentrations of CO_2. When radiative forcing is evaluated as positive, the energy of the Earth–atmosphere system will ultimately increase, leading to a warming of the system. In contrast, for a negative radiative forcing, the energy will ultimately decrease leading to a cooling.

Important challenges for climate scientists are to identify all the factors (not just CO_2) that affect climate. However, it is important to note that modern mainstream climate science and the IPCC do not generally consider water vapour a greenhouse gas, which is why it is not shown in Figure 19.3. This is on the basis that it is not a constant component of the atmosphere. This is a somewhat unique and particular approach – and not consistent with the view of many analytical chemists who are experts in modern spectroscopy, or even with Professor Arrhenius. Water vapour is, of course, incorporated into the IPCC models, but not as a greenhouse gas that causes radiative forcing as such.

Figure 19.3 shows the various contributions to radiative forcing according to the IPCC (IPCC 2007) and also the US EPA (EPA 2016). The most important forcings, according to this IPCC model, are attributed to the greenhouse gases, particularly CO_2. According to the IPCC, forcings due to human-caused sources far outweigh any natural factors, for example changes in the Sun's irradiance.

Figure 19.3 Radiative forcing according to the IPCC

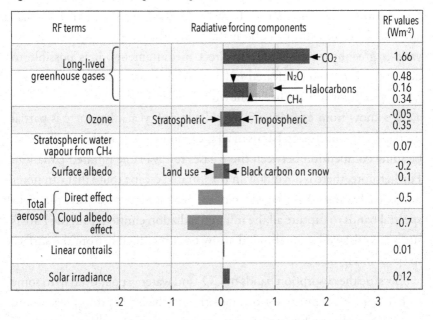

Water vapour in the lower troposphere is not recognized as a radiative forcing component by the IPCC.

Source: IPCC 2007.

While the simplified version of the greenhouse effect presented in Figure 19.1 is generally accepted, as is the nature of the electromagnetic spectrum in Figure 19.2, there is less agreement about the magnitude of various contributing factors to global warming, specifically through radiative forcing as shown in Figure 19.3.

Scientists have cast doubt on the levels of forcing from each of the components listed in Figure 19.3, and on the contribution of CO_2 to the greenhouse effect (Christy et al. 2010; Douglass & Christy 2009; Lindzen 1981; Lindzen et al. 2001). In short, Figure 19.3 is highly contested. Of particular concern, Figure 19.3 does not include water vapour, which is considered by many to be the most important greenhouse gas.

Quantifying the impact of carbon dioxide

Since the crude calculations of Professor Arrhenius more than 120 years ago, much more is known about the infrared absorption spectra of the various greenhouse gases. From direct measurements, it is possible to determine how much infrared radiation will be absorbed and transmitted at specific wavelengths, as shown in Figure 19.4.

We know from experiments over the last 40 years that there is partial overlap in the absorption bands of all greenhouse gases, so there will always be some competition between these gases for available infrared radiation. Furthermore, there is a window in the infrared escape route that can never be closed because there are no natural greenhouse gases with the right spectral bands to capture all the infrared radiation emitted from the Earth's surface, as shown by the thin red arrow pointing directly from the surface to space in Figure 19.1.

Because the absorption bands of CO_2 and water vapour overlap to some extent they will compete for this radiation in the lower atmosphere – and so the effect of a doubling of the concentration of CO_2 will vary according to the concentration of water vapour. While the concentration of CO_2 is relatively constant in the lower atmosphere, there is tremendous variability in concentrations of water vapour. Indeed, the concentration of water vapour varies with altitude, latitude and land use. Therefore, calculating the effect of a doubling of atmospheric CO_2 is complicated.

Dr Barrett's research at Imperial College in the 1980s revealed that nearly all the radiation from the ground – the wavelengths of which can be absorbed by spectral lines comprising the CO_2 bands – is locked into the air sample within a distance of 100 m from the ground. Considering just the lowest 100 m of the atmosphere – where the effect of increasing CO_2e would theoretically be most pronounced – a doubling of CO_2, according to Dr Barrett's calculations will result in an increase of absorption by CO_2 of just 1.5%, as shown in Table 19.1.

Figure 19.4 Infrared spectra of dominant greenhouse gases

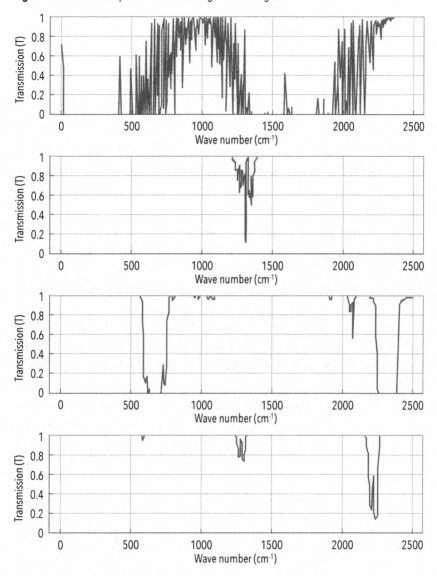

Infrared spectra of the four dominant greenhouse gases at sea level: water vapour (top), methane (second), carbon dioxide (third) and dinitrogen monoxide (bottom). If T=0, all the radiation is absorbed; if none is absorbed then T=1.

Source: Barrett 2005.

Table 19.1 Contributions to the absorption of the Earth's radiance by the first 100 metres of the atmosphere

GHG	% Absorption	Absorption relative to water vapour = 1
Water vapour	68.2	1.000
CO_2 (285 ppmv)	17.0	0.249
CO_2 (570 ppmv)	18.5	0.271
CH_4	1.2	0.018
N_2O	0.5	0.007
Total [water, CO_2 (285 ppmv), CH_4, N_2O]	86.9	
Combination with 285 ppmv CO_2	72.9	1.069
Combination with 570 ppmv CO_2	73.4	1.076

Source: Barrett 2005.

This would suggest that the values calculated by Professor Arrhenius and included in the IPCC General Circulation Models are much too high. Similar conclusions have been drawn by others with expertise in spectroscopy.

Like Dr Barrett, Professor Alfred Laubereau is an expert in spectroscopy. Now retired, he is a German physicist with more than 200 scientific articles in the area of spectroscopy to his name, written over a period of more than 40 years while at the physics department at the University of Munich.

In 2013, he published a paper reporting on the likely direct impact of rising CO_2 concentrations to global warming. This investigation essentially reported on spectroscopic measurements of pure CO_2 in the infrared region using a cell of path length 10 cm with temperatures in the range 220 K to 295 K based on a five-layer model for the greenhouse effect – including a surface layer and four atmospheric layers

Using the reported increase in atmospheric CO_2 concentration from 290 to 385 ppm between 1880 and 2010, Professor Laubereau and his coauthor Hristo Iglev derived a direct temperature rise attributable to CO_2 of about 0.26 °C. Even including the simultaneous feedback effect of atmospheric water vapour, they nevertheless determined that CO_2 contributed somewhat less than 33% of its reported contribution to global warming (Laubereau & Iglev 2013).

This study, like the early work of Barrett's, represents one of very few published studies detailing the results from actual experimental work involving spectroscopic measurements. More typically, values are extrapolations from General Circulation Models based on the crude calculations of Professor Arrhenius.

Working independently of Professors Laubereau, and Barrett, Dr Douglas Lightfoot and Dr Orval Mamer have also undertaken detailed experiments to calculate atmospheric radiative forcing of CO_2 at different concentrations (Lightfoot & Mamer 2014). These two Canadian scientists are very critical of the use of the Beer–Lambert law in the General Circulation Models relied upon by the IPCC to determine radiative forcing from CO_2. As Lightfoot and Mamer explain, there is no theoretical basis for using this relationship, because water vapour and CO_2 compete unevenly over the wavelength absorption range of infrared radiation.

Lightfoot and Mamer detail two different methods for calculating the effect of CO_2 on global temperatures. The first approach, based on experimentation, derives a quadratic model that gives the same change in radiative forcing as the IPCC's logarithmic model over the range 275 to 378 ppm CO_2. However, outside this range, their quadratic model gives a better fit to the experimental results, and at 378 ppm indicates a radiative forcing of 8.67 Wm^{-2}, and a change in radiative forcing of 3.26 Wm^{-2} for the doubling of CO_2 from 275 to 550 ppm.

These are both larger in value than that shown in Figure 19.2 for CO_2. Then again, Drs Lightfoot and Mamer indicate that the total contribution of CO_2 to radiative forcing, relative to all the greenhouse gases shown in Figure 19.2, is just 2.7%.

Drs Lightfoot, Mamer and Barrett emphasise the contribution of water vapour as a greenhouse gas, with Lightfoot and Mamer suggesting it contributes 96% to current greenhouse-gas warming. The IPCC models do not include water vapour as a forcing, but do include the radiative effects of water vapour with feedback due to the higher temperatures from elevated levels of atmospheric CO_2, which allow a larger water vapour content thereby increasing warming further.

The second method for estimating radiative forcing employed by Lightfoot and Mamer uses worldwide hourly measurements of atmospheric temperature and relative humidity. Again, they conclude that CO_2 is responsible for approximately 2.7% of the total radiative forcing of all of the greenhouse gases, and specifically conclude that a doubling of CO_2 would result in an increase in the Earth's temperature of 0.33 °C.

This value of 0.33 °C is similar to Laubereau's value of 0.26 °C – and an order of magnitude less than values currently inserted into the General Circulation Models that are used to influence global energy policy and emissions limits.

Conclusion

The IPCC has published many detailed assessment reports with specific estimates of the contribution of CO_2 to global warming. While their first assessment report published in 1991 predated the work of Laubereau, Iglev, Lightfoot and Mamer, it should have considered the finding of Dr Barrett, and other earlier twentieth-century scientists with expertise in spectroscopy. Instead, simply ignoring this body of research which built on and refined the work of Joseph Fourier and Svante Arrhenius, the IPCC choose to extrapolate from Arrhenius's theoretical calculations – made in part through comparisons of the atmosphere of the Earth with that of the Moon. The original 'high estimate' by the IPCC, consistent with Arrhenius's speculative theory, suggests a temperature increase from CO_2 of approximately 6 °C, as shown in Figure 19.5.

Figure 19.5 Temperature rise due to greenhouse gases

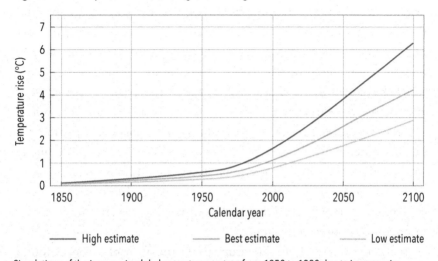

Simulations of the increase in global mean temperature from 1850 to 1990 due to increases in greenhouse gases (excluding water vapour), and predictions of the rise between 1990 and 2100 as determined by the IPCC in its first assessment report.

Source: Houghton et al. 1991.

Meanwhile the detailed spectroscopy undertaken by Barrett and other contemporary spectroscopy experts, which provides the basis for much more accurate calculations than those performed by Arrhenius, has been ignored by the IPCC.

Somewhat hypocritically, advocates of IPCC science, for example Martin Lack in his book *The Denial of Science* (2013), dismiss out-of-hand Barrett's work on the basis that he is a chemist – not a climate scientist. At the same-time they laud the work of Professor Arrhenius who won the 1903 Nobel prize for chemistry.

Techniques from chemistry, particularly modern spectroscopy, will be critical to improving our understanding of the physics of carbon dioxide in the Earth's atmosphere. Furthermore, this is at the core of IPCC attempts to understand the science.

Back in 1896, the chemist Arrhenius acknowledged the need for detailed experimentation, while indicating that the necessary 'apparatus' were not at his disposal – so he was reduced to *indirect* calculations in order to determine the likely impact of a doubling of atmospheric levels of CO_2.

It was not until the 1950s that modern spectroscopy techniques, which facilitated the later work of Barrett, Laubereau, Iglev, Lightfoot and Mamer, were developed. What is now needed is for the calculations of these spectroscopy experts to be tested in controlled experiments. Furthermore, it would not be difficult to build an enclosed, highly controlled and monitored experimental system at least 100 m in length, where the predictions of Barrett could be tested. The cost of such an experimental system would be small compared to the enormous expenditure on computer modelling.

A recent review of the literature using the Web of Science[3] found just one such reported investigation from a Japanese researcher using an enclosed system 2 m in length (Kawamura 2016). In contrast, a similar search gives over 12,000 scientific papers on the topic of General Circulation Models and climate. There is a place for modelling, but it must be informed by observational data.

Mr Pruitt, the new head of the US EPA, was correct to insist there is a need for more analysis and review. Given the importance of knowing the precise relationships between greenhouse gases and climate change, his administration might also consider funding research where actual physical measurements are made to test theoretical predictions – most recently the predictions based on the detailed measurements of the absorption and transmission of infrared radiation by CO_2 and other greenhouse gases in the laboratory, such as, Barrett (2005) Lightfoot and Mamer (2014), and Laubereau and Iglev (2013).

---

3   webofknowledge.com

20 Carbon Dioxide and the Evolution of the Earth's Atmosphere

Dr Ian Plimer

Climate change has taken place for thousands of millions of years.

Climate change occurred before humans evolved on Earth. Any extraordinary claim, such as that humans cause climate change, must be supported by similarly extraordinary evidence, but this has not been done. It has not been shown that any measured modern climate change is any different from past climate changes. The rate of temperature change, sea-level rise, and biota turnover is no different from the past.

In the past, climate has changed due to numerous processes, and these processes are still driving it. During the time that humans have been on Earth there has been no correlation between temperature change and human emissions of carbon dioxide (CO_2). Past global warmings have not been driven by an increase in atmospheric CO_2.

Without correlation, there can be no causation.

The underpinning assumption

The underpinning assumption is that human emissions of CO_2 drive global warming and, in order to arrest the warming trend, human emissions of CO_2 must therefore be reduced. But it has yet to be shown that the human emissions of CO_2 actually do drive global warming, despite decades of generous research funding and hundreds of years of science.

Emissions of CO_2 from human activities – such as the production of metals, energy and cement; land, air and sea transport; and heating and cooling – account for about 3% of total annual emissions.

If it could be shown that the human emissions of CO_2 do drive global warming, then it would also have to be shown that natural emissions of CO_2, which amount to 97% of the total annual emissions, do not drive global warming. But this has not been done. In fact, there is very strong evidence from many disciplines of science, especially geology, to show that over thousands of millions of years, neither human, nor natural CO_2 emissions have driven global warming.

Natural emissions are mainly from oceanic degassing and mantle degassing via millions of submarine volcanoes, mid-ocean ridge volcanicity and ocean floor fractures with less voluminous emissions from mantle degassing via terrestrial volcanoes, mountain degassing, fractures, earthquakes, respiration, and decomposition of biota.

In past times, the Earth's atmosphere had many times the current CO_2 level, see Figure 20.1. Yet there was no runaway greenhouse effect, no irreversible warming, and no catastrophe. Climate models, and research and energy policy, are underpinned by the fallacious assumption that human emissions of CO_2 drive global warming. This erroneous assumption has been made in the absence of considering the effects of the Sun, tectonics, the Earth's orbital oscillations, oceanic oscillations, and extraterrestrial cosmic radiation.

The oceans remove most of the CO_2 from the atmosphere. What is rarely considered is that: CO_2 is plant food, it is rapidly removed from the atmosphere, and it has an atmospheric life of about seven years. Soils and rock weathering also quickly remove CO_2 from the atmosphere. There is an underlying assumption that once humans emit a CO_2 molecule, it stays in the atmosphere for a very long time, or even forever. Any CO_2 emitted by humans is part of the carbon cycle that involves recycling CO_2 through the atmosphere, life, water, and

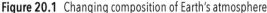

Figure 20.1 Changing composition of Earth's atmosphere

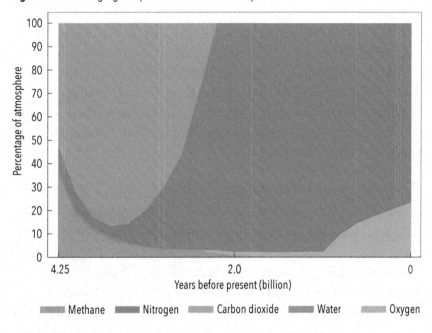

The fraction of total atmospheric mass by that constituent.

Source: adapted by permission from Elsevier – Hart, MH, 'The evolution of the atmosphere of the Earth', *ICARUS*, vol. 33, pp.23-39, copyright 1978.

rocks. Carbon cycles have been in operation for thousands of millions of years.

The human-induced global warming ideology is underpinned by the perception that the planet is static and that dynamic change only occurred once humans started to emit CO_2. Nothing could be further from the truth. The Industrial Revolution in Europe and the US triggered a great increase in human emissions of CO_2 from the use of fossil fuels; while another industrial revolution, which is taking place today – in Asia – is causing the rate of emissions of CO_2 from the use of fossil fuels to accelerate; however, there is no correlation between the emissions of CO_2 by humans and temperature.

If it is only the human emissions of CO_2 that drive global warming, then there should have been no warming and cooling events before the Industrial Revolution. However, pre-Industrial Revolution climate changes were fast and greater than any change measured today.

Natural variations in CO_2 over time have not driven climate change. In fact, it is the opposite. Natural variations in climate change have had a delayed effect on atmospheric CO_2.

Deep time, space and carbon dioxide

Greenhouse gases, such as water vapour (H_2O), methane (CH_4) and CO_2, are common gases in our solar system. Even before the formation of our solar system, H_2O, CO_2, CH_4, and other more complex carbon compounds such as alcohols, were present in the universe.

Comets contain H_2O and CO_2 ice; liquid CH_4 and liquid CO_2 are present in and on other planets and their moons; CO_2 is ubiquitous in planets, moons and their ice caps; and CO_2 is a common and commonly abundant component of the atmosphere of other planets and their moons. Protoplanetary bodies and fragmented protoplanets (for example, asteroids and meteors) contain these greenhouse gases, as well as CO_2-bearing carbonate minerals. Recent discoveries of H_2O-, CO_2- and bicarbonate-bearing minerals on Mars and in Martian meteorites have created interest in the possibility that there may have been life on early Mars.

Yet, on Earth, the US Environmental Protection Agency (EPA) has deemed that a ubiquitous and abundant gas in the solar system (CO_2) is a pollutant on just one of the solar system's planets, Earth.

Planet Earth and carbon dioxide

Planet Earth has been degassing since its formation 4567 million years ago. Some gases were retained in the atmosphere (for example, nitrogen, noble gases such as argon); other gases, such as H_2O, condensed to form oceans, which initiated a water cycle. Yet other gases, such as sulphurous gases, CO_2 and CH_4, were concentrated in the early atmosphere and later these were recycled and sequestered.

Until photosynthetic life evolved in the Earth's middle age, there was no oxygen gas (O_2) in the atmosphere. Various common processes, such as lightning, convert O_2 to ozone (O_3). It is the presence of this O_3 in the Earth's atmosphere that would inform any hypothetical extra-terrestrials that there is life on Earth, not the electromagnetic signals that we emit from radio and television.

Hundreds of planets associated with other stars, and so outside our solar system, have now been recognised. Judging by the absorption of starlight by CO_2 we can determine that this is a very common planetary gas. However, no O_3 has been detected, showing that photosynthetic life outside the solar system has not yet been discovered.

The Earth's crust and its atmosphere have been evolving – with the constant recycling of the oceans, soils, life, sediments and rocks – for thousands of millions of years. Early Earth had a CO_2-rich atmosphere, possibly with as much as 20% by volume – compared to the 0.04% by volume today, as shown in Figure 20.1.

Planet Earth is remarkable because liquid water has existed on the planet for some 4000 million years suggesting a fixed orbital address and a very stable star. Despite great changes in atmospheric chemistry, it is the long-term presence of liquid water on Earth that has allowed life to evolve.

If planet Earth was outside the habitable zone in our solar system (like, for example, Venus or Neptune) or had lost its magnetic field (as has Mars), then Earth could have had permanent ice cover or lost the oceans by evaporation and solar wind stripping.

During the time of life on Earth, the atmosphere has changed from at least 5% (and perhaps 20%) CO_2 to the present atmosphere with

0.04% CO_2, yet life evolved, ice ages came and went and the continents moved, pulling apart, and stitching back together. There was no irreversible catastrophic global warming in the geological past, despite the Earth's CO_2-rich atmosphere.

If a high atmospheric CO_2 in the past did not drive global warming, then a far lower atmospheric CO_2 in the modern atmosphere cannot drive global warming. Atmospheric CO_2 is one of the many, but very minor, reasons why the Earth's surface temperature is habitable.

If atmospheric CO_2 drives global warming, why didn't the planet have runaway global warming in the geological past? We know, for example, that there were a number of major long-lived ice ages during times of very high atmospheric CO_2 (Pangolian, Huronian and Cryogenian), when, for at least two of these ice ages, kilometre-thick ice sheets were at sea level and at the Equator. With the planet either a snowball, or slush ball, sea level changed vertically by many kilometres. The origin of these greatest climate changes that the Earth has ever experienced is still being hotly debated.

While some areas were covered by ice during a glaciation event in an ice age, areas not covered by ice were arid. This is seen time and again in the geological record with 'red beds' (red desert sands) and evaporites (rock salts) occurring coevally with glacial debris. In the geological record, deserts occurred in cold times and not warm times. However, climate catastrophists promote the ideology that deserts will increase in size during a hypothetical global warming.

In many places, debris left behind by retreating ice in the Cryogenian was covered by fossiliferous limey rocks known as cap carbonates. The ice sheets would have been present when the temperature was <0 °C and more likely <40 °C. The chemical fingerprint of the cap carbonates show that they formed at a temperature of >40 °C. This shows that past temperature changes were huge and rapid and anything measured today pales into insignificance.

Although there are many theories on the origin of ice ages, it is still not known why the planet dropped into icehouse conditions, and why the planet changed from icehouse to greenhouse conditions. If we cannot understand the reason, or reasons, for the greatest climate changes in the past, then we need to be cautious about the very slight modern climate changes.

The weight of ice in modern ice sheets has created crustal stretching and subsidence in Greenland and Antarctica. Removal of ice over the last 12,000 years during the current interglacial period has resulted in the rising of land in Scandinavia, the British Isles and North America. The bulk of the Greenland and Antarctic ice sits in basins. Ice is always moving and a polar ice sheet or glacier is not static, or forever. Ice flows uphill and out of these basins by pressure and then flows downhill in glaciers by pressure. Pressure on ice results in constant recrystallisation and ice movement. Air temperature can hasten surficial ice melting, whereas pressure and melt waters at the base of a glacier facilitate movement and melting.

There are numerous volcanoes, hot rocks and steam vents beneath the Antarctic ice sheet. These melt ice. The Gakkal Rise in the Arctic Ocean contains active, explosive volcanoes that add heat to the Arctic waters, which can then melt terrestrial and sea ice. An ingress of warmer water into the Arctic during cyclical lunar tidal activity has resulted in a cyclical reduction of sea ice. As well, some rocks beneath the Greenland ice sheet are highly radioactive and the breakdown of radioactive elements produces heat that can melt ice.

Because of a 130 m sea-level rise over the last 12,000 years during the current interglacial, some of the Antarctic Ice Sheet is now pinned to islands and underlain by water. The substrate of warmer sea water can induce ice melting. Glaciers flow downhill and are replenished by precipitation.

However, a contracting ice sheet or a retreating glacier does not necessarily indicate global warming of the atmosphere, especially as air

303

has a low heat capacity compared to water. Furthermore, if air temperature rises it does not necessarily mean that ice melts – a great deal of heat needs to be added to ice at 0 °C before it melts. There are numerous reasons why ice expands and contracts and, at present, ice sheets, sea ice and some glaciers are expanding whereas others are contracting.

For half of geological time, carbonate rocks did not form from sediments or oceanic precipitates. A huge drawdown of CO_2 from the atmosphere via oceans into carbonate rocks occurred in the Proterozoic (2500 to 583 million years ago) during these three ice-age events. Massive volumes of dolomite were precipitated. Dolomite ($CaMg[CO_3]_2$) contains 48% CO_2 by weight.

Experimental studies show that dolomite can only be precipitated from ocean waters when there is a very high quantity of CO_2 in the atmosphere. When the CO_2 content is lower, calcite ($CaCO_3$; 44% CO_2 by weight) is precipitated, a process that is taking place at present. When the atmospheric CO_2 content is extremely low, plants will die (that is, at <200 parts per million volume, or ppmv, atmospheric CO_2) and gypsum ($CaSO_4.2H_2O$) will precipitate from the oceans, a process that is not taking place at present, nor has ever taken place. The volume of Proterozoic rocks can be measured to back-calculate the atmospheric CO_2 content at that time.

During the Phanerozoic (last 583 million years ago), the atmospheric CO_2 content has varied greatly. For the last half billion years the global atmospheric CO_2 content has been decreasing and, at present, the atmospheric CO_2 content is relatively low. With the rise in animals with shells, terrestrial vegetation, and the deposition of fossiliferous limestone containing calcite ($CaCO_3$), the decrease in atmospheric CO_2 has accelerated with significant decreases during the times of expansion of the plant kingdom.

Atmospheric CO_2 was up to least ten times higher than at present; four Phanerozoic ice ages (Ordovician, Permo-Carboniferous, Jurassic and Quaternary) were initiated when atmospheric CO_2 was higher than the present. These four Phanerozoic ice-age events were coeval with desertification at lower latitudes. The hypothesis that an atmosphere with a slightly higher CO_2 content than at present will drive global warming is therefore invalid. Furthermore, there is a lack of correlation between atmospheric CO_2 and temperature.

The past is the key to the present, and geology consistently shows that the assumptions made by the global warming idealists are incorrect.

Conclusion

Geology shows us that there is no relationship between atmospheric CO_2 and climate. If there is no correlation between atmospheric CO_2 in past times when the atmosphere had a far higher CO_2 content, then there are no reasons why a minute increase in atmospheric CO_2 in today's atmosphere, which is already depleted in CO_2, should drive climate. Larger processes have driven climate change and still do.

The past shows us that every single catastrophic event predicted by climate activists is wrong (for example, rate of temperature rise, influence of CO_2 on climate, sea-level rise, extinction, destruction of coral reefs, ocean 'acidity', atmospheric heating of oceans). It is no wonder that climate catastrophists avoid debating science with geologists.

For a scientific hypothesis to be validated, it must be in accord with all data. This is the coherence criterion of science. Because the hypothesis that human emissions of CO_2 drive global warming is not in accord with what the geological record has revealed, then the hypothesis is invalid.

21 The Geological Context of Natural Climate Change

Dr Bob Carter[1]

The issue of dangerous human-caused global warming is a complex one. It can be assessed meaningfully only against our knowledge of natural climate change, which is incomplete and in some regards even rudimentary. The geological record of climate reveals many instances of natural changes of a speed and magnitude that would be hazardous to human life and economic well-being should they be revisited upon our planet today. Many of these changes are unpredictable, even in hindsight. That such natural changes will occur again in the future, including both episodic step events and longer-term cooling and warming trends, is certain.

The geological record of climate change

The focus of the Intergovernmental Panel on Climate Change (IPCC) activity has been on erecting computer models that are rooted in a knowledge of the last 150 years of instrumented climate observations, sometimes extending back to around 1000 years, as defined by proxy measurements such as tree-ring analysis. This is an utterly inadequate period over which to seek to understand climate change.

---

1   Adapted from chapter 1 of *Climate: The Counter Consensus*

Using climate records that represent the last several million years, palaeoclimatologists and palaeoceanographers have established a sound understanding of the natural patterns and some of the mechanisms of climate change. The most important evidence comes from sediment cores from beneath the deep-sea floor and ice cores through the Greenland and Antarctic ice caps. Any one such core does not, of course, depict global climate; however, suitable cores yield climate data that are representative of a wide region and some may even approximate a global pattern. Generally agreed inferences from these data are as follows.

Milankovitch frequences of climatic change

Between about six and 3.5 million years ago, during the period that geologists term the Pliocene Epoch, small warm–cold climatic oscillations occurred around a mean temperature that was 2–3 °C warmer than today, as shown in Figure 21.1. Starting 3.5 million years ago, global temperature embarked on a steady decline, towards the end of which the background 41,000-year climatic cycles became overprinted with more severe, 100,000-year, glacial–interglacial oscillations; and also a 20,000-year oscillation superimposed on the other two.

These three fundamental frequencies of climatic oscillation are termed Milankovitch frequencies (Graham 2000; Hays et al. 1976), after Serbian geophysicist Milutin Milankovitch who, early in the twentieth century, spent almost twenty years laboriously calculating the first handmade graphs of Earth's recent climatic history – graphs that can now be produced more accurately on a laptop computer in seconds. Milankovitch's insight – following that of the self-taught British geologist and physicist John Croll, who published the important and novel book *Climate and Time* in 1875 – was to appreciate that the distribution of radiant solar energy received across planet Earth changes through time in correspondence with fluctuations in the Earth's orbit. These orbital variations are caused by gravitational interaction with the other planets

Figure 21.1 Pacific Ocean deep-water temperature

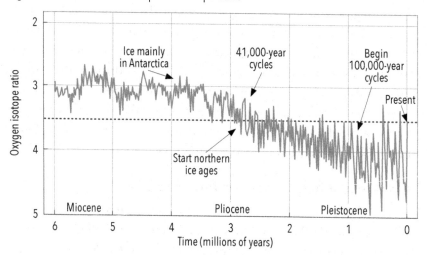

Composite curve from Ocean Drilling Program (ODP) Sites 846 and 849, Equatorial Pacific, based upon oxygen isotope ratios measured in benthic foraminifera in marine cores. The dotted red line indicates today's temperature. As the oxygen isotope ratio decreases the temperature increases.

of the solar system, and affect both the tilt of the Earth's axis and the shape of its orbit around the Sun. Specifically, the path of the orbit varies from more to less elliptical on a 100,000-year scale; the tilt of the Earth's axis varies slightly, between about 22.1° and 24.5° on a 41,000-year cycle; and, the Earth's tilted axis also precesses ('wobbles') on a roughly 20,000-year cycle. The changing geometries exert a marked effect on Earth's seasonality, which in turn controls the accumulation of snow and ice at the high latitudes at which the great Northern Hemisphere ice sheets accumulated over the last few million years.

Since about 0.6 million years ago, after what has been termed the mid-Pleistocene (climatic) revolution, each major glacial–interglacial oscillation has occurred on the longer, 100,000-year periodicity. For more than 90% of this time the Earth's mean temperature was cooler, and often much cooler (up to ~6 °C), than today. Warm interglacial periods comprised less than 10% of the time, and on average lasted only 10,000 years. Civilisation and our modern society developed during the most recent warm interglacial period (the Holocene), which has already lasted 10,000 years. In many places, temperatures earlier in the Holocene – during the 10,000- to 9000-year period that corresponded to a maximum in total solar insolation at 65°N (Berger & Loutre 1991), formerly referred to as the climatic optimum – were 1 °C to 2 °C warmer than today (Davis et al. 2003; Williams et al. 2004; Kaufman et al. 2004).

The spectacular Milankovitch climatic record was first captured in deep-ocean sediment cores (Emiliani 1955; Shackleton & Opdyke 1973), and later confirmed by studies of glacial ice cores from Greenland and Antarctica (Dansgaard et al. 1969; Watanabe et al. 2003; Augustin et al. 2004). Excitingly, these latter cores have the unique capability of yielding measurements of ancient atmospheric chemistry from air samples that are captured as bubbles within the ice. Measurements on the gas contained in the bubbles demonstrate that changes in past temperature and atmospheric carbon dioxide (CO_2) concentration in such cores occur in close parallelism. In detail, however, the changes in temperature precede their parallel changes in CO_2 by between ~800 and 2000 years (Mudelsee 2001). This vital point establishes that CO_2 cannot be the primary forcing agent for temperature change at the glacial–interglacial scale.

Abrupt climatic events

The recent climate record is not only cyclic, it is also punctuated by episodes of abrupt climate change, when climate sometimes traversed almost its full glacial–interglacial range in a period as short as a few years to a few decades. A well-known and dramatic example of this concerns a post-last-glacial warming in the Northern Hemisphere, where Greenland ice cores and other records show that abrupt warming to almost full interglacial level 14,500 years ago was followed by a reversion to temperatures of glacial intensity termed the Younger Dryas episode, which in turn was followed by a sharp resumption of warming 11,600 years ago that continued more gently up to the start of the Holocene, as shown in Figure 21.2. The two episodes of sharp warming occurred in a three-year and 60-year period, respectively, while the cooling into the Younger Dryas stretched over 1500 years, after which its deepest cold phase continued for another 1000 years (Taylor et al. 1993; Steffensen et al. 2008; Brauer et al. 2008; Lie & Paasche 2006). The causes of such abrupt climate changes as these remain largely unknown – though changes in ocean currents, changes in the delivery of fresh water to the

Figure 21.2 Surface air temperatures since the last glacial period

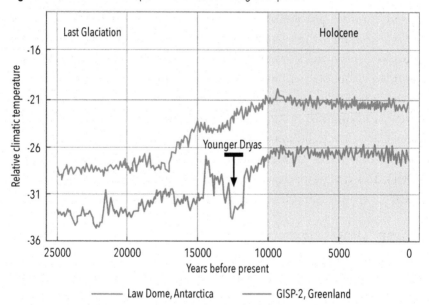

Antarctica (Law Dome ice core) and Greenland (GISP-2 ice core) temperatures based on ice core measurements of deuterium and oxygen isotopes, respectively. Note the conspicuous climatic cooling event called the Younger Dryas in Greenland, and that the last 10,000 years (the Holocene) comprises a period of gentle cooling that continues today.

ocean (including by the sudden drainage of giant glacial lakes), changes in ocean current flow, and even extraterrestrial impact, remain among the candidate explanations.

Younger Dryas summers and winters are estimated to have been 4 °C and 28 °C cooler than today's, respectively. A computer-model reconstruction of sea-ice extent during Younger Dryas winters projects that the ice advanced to cover nearly all of the North Atlantic Ocean north of a latitude of 40°N (Broecker 2006), which corresponds approximately to a line connecting the Brittany peninsula (France) and northern Newfoundland (Canada). Apparently, very little commercial shipping would have been coming out of Europe's Channel ports. Noting that ice caps take many thousand years to grow or melt, but that sea ice can be built in a year or two, one wonders idly what contingency plans have been laid by northern Atlantic nations for the recurrence of a sharp climatic cooling of the Younger Dryas type – with attendant dislocation of much transatlantic marine commerce by sea ice, and an increased demand for energy for heating due to the intense cold. The last time that I looked, however, European and US governments seemed to be concerned only with *l'affaire du jour* of global warming.

Temperatures during the Holocene

Twenty thousand years ago, the peak of the last great glaciation, is but a geological heartbeat away. Two hominid species inhabited our planet during the earlier parts of this ice age: *Homo neanderthalensis* and *Homo sapiens*. Life must have been tough, caves very welcome, and the ability to light fires an absolute blessing. The climatic warming that then ensued was punctuated by the Younger Dryas, and ended 10,000 years ago, at the dawn of the Holocene. Starting around 12,000 years ago, during the cultural period called the Mesolithic, *Homo sapiens* discovered how to make pottery, grow farm crops, domesticate animals, smelt first bronze and then iron, and how to develop city civilisations – many of these developments surely being aided by the relative warmth of the climate.

Compared with the vicissitudes of the Younger Dryas, the Holocene climate may appear to have been relatively stable, but nonetheless, and as always, change was a constant. First, the long-term temperature record of the Holocene is one of cooling by 1–2 °C from the post-glacial climatic optimum in the early Holocene (Figure 21.2). Second, throughout the Holocene, a 1500-year long climate cycle – with a similar magnitude of 1–2 °C and sometimes called the Bond Cycle – was conspicuous (Bond et al. 2001; Skilbeck et al. 2005; Moros et al. 2009; Willard et al. 2005; Avery & Singer 2008). In Greenland, the three most recent historic warm peaks of the Bond Cycle (Minoan, Roman and Medieval Warm Periods) all attained or exceeded the magnitude of the late twentieth-century warming. A variety of detailed proxies for past temperature show a similar climatic pattern in many localities around the world (Soon & Baliunas 2003; Idso 2010), and some of the best of these records have been used to construct a global temperature estimate for the last 1500 years that confirms the greater warmth of the Medieval over the late twentieth-century warm period (as shown in Figure 21.3).

Is modern climate change unusual?

Importantly, compared with the ancient climate record, temperatures during the late twentieth century were neither particularly high nor particularly fast-changing. Not only were recent Bond Cycle peaks a degree or so warmer than modern temperatures, but temperatures in Antarctica for the three interglacial periods that preceded the Holocene were up to 5 °C warmer than today, as shown in Figure 21.4, and temperatures ~2–3 °C warmer probably characterised much of the planet during the Pliocene (Figure 21.1). Meanwhile, as to the rate of temperature change, analysis shows that the two twentieth-century warming pulses occurred at a rate of ~1.5 °C per century, which falls well within the natural rates of Holocene warming and cooling exhibited by high-quality records like the Greenland ice core (Greenland Ice Core Project, Figure 21.4).

Figure 21.3 Global temperature for the last 2000 years

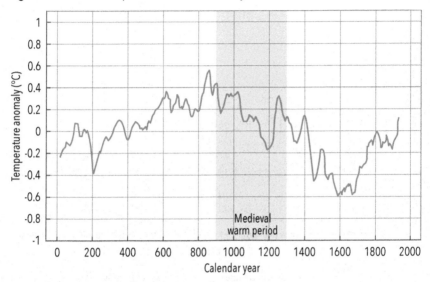

The series has been reconstructed from eighteen sets of geological proxy measurements, including pollen and diatoms from lake cores, and isotope data from ice cores and speleothems. Note the Medieval Warm Period was significantly warmer than the twentieth century up until 1980.

This rate of change compares with a twentieth-century warming rate of ~1 °C per century for Greenland, based on oxygen isotope measurements in the GISP2 ice core.

It is clear from these various facts that a warmer or cooler planet than today's is far from unusual, and that nature recognises nothing 'ideal' about mid-twentieth-century temperatures. It is also clear that climate changes naturally all the time. The idea implicit in much public discussion of the global warming issue, that climate was stable (or constant) prior to the Industrial Revolution, after which human emissions have rendered it unstable, is fanciful. Change is simply what climate does.

Figure 21.4 Rate of temperature change over the Greenland ice cap

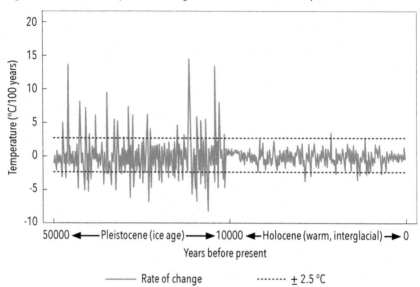

The rate of change during the Holocene generally fluctuated between ±2.5 °C per century of both warmings and coolings (between the two horizontal dotted red lines). Note the horizontal axis is not to a linear scale.

Climatic cycles of solar origin

The Earth's surficial skin, within which we live and travel, has two main sources of heat energy. Some heat comes from deep within the planet, derived by conduction from the molten core and the decay of radioactive elements in the mantle and crust, in the very small amounts of 20 to 40 milliwatts/square metre (mW/m^2), depending upon whether it is measured above oceanic (hotter) or continental (cooler) crust (Sclater et al. 1980). But the dominant energy source by far is the Sun, whose direct radiation provides ~342 watts/square metre (W/m^2) of heat at the top of the atmosphere. It is this solar heat, or more strictly its redistribution by radiation and convection, that drives the climate system.

It is long known that the Sun's direct radiation of light and heat (which scientists term the total solar irradiance, or TSI) varies slightly on the scale of the approximately eleven-year sunspot cycle. During this cycle, the Sun builds up from a small to a large number of spots, which then decline in number again. Superimposed on top of the sunspot cyclicity during historic times has been a long-term trend of slight increase in overall irradiance, which amounts to approximately 2 W/m^2 since the seventeenth century. Supporters of the IPCC, noting that this increase represents only 0.1% of the total solar irradiance, are fond of claiming that this small change is inadequate to explain the increases observed in global temperature since the nineteenth century. But the main way in which the changes in the Sun's irradiance affect Earth's climate is through their direct influence on seasonal and annual climate cycles, which is a much more important effect than the minor and indirect influence of increasing atmospheric CO_2. Also, to restrict one's attention to changes in irradiance only – which the IPCC does in arguing that because the Sun exhibits only minor long-term irradiance change it can have but little effect on climate – is to throw the baby out with the bathwater in spectacular fashion, for in reality there are many other important ways in which solar variations affect Earth's climate.

The other aspects of Earth–Sun energy interrelations that are known to play a role in climate include the following:

- Variations in the intensity of the Sun's magnetic fields on cycles that include the Schwabe (eleven-year) (Meehl et al. 2009; Soon et al. 1996; Soon 2005); Hale (22-year) (Alexander 2005); and Gleissberg (70- to 90-year) (Lassen 2009; Cini Castagnoli 2005) periodicities.
- The effect of the Sun's plasma and electromagnetic fields on rates of Earth rotation, and, therefore, the length of day (LOD) (Lambeck & Cazenave 1976).
- The effect of the Sun's gravitational field through the 18.6-year Lunar Nodal Cycle – which causes variations in atmospheric pressure,

temperature, rainfall, sea level and ocean temperature, especially at high latitudes (Yasuda 2009).

- The known links between solar activity and monsoonal activity (Agnihotri & Dutta 2003; Liu et al. 2009), or the phases of climate oscillations – such as the Atlantic Multidecadal Oscillation, a 60-year cycle during which sea-surface temperature varies ~0.2 °C above and below the long-term average, with concomitant effects on Northern Hemisphere air temperature, rainfall and drought.

- Magnetic fields associated with solar flares, which modulate galactic cosmic ray input into the Earth's atmosphere, and in turn may cause variations in the nucleation of low-level clouds at up to a height of a few kilometres. This causes cooling, a 1% variation in low cloud cover producing a similar change in forcing (~4 W/m²) as the estimated increase caused by human greenhouse gases. This possible mechanism is controversial and remains under test in current experiments devised by Henrik Svensmark (Svensmark et al. 2007) at the European Organization for Nuclear Research (CERN). But irrespective of the results of these experimental tests, and of the precise causal mechanism, Neff et al. (2001) have provided incontrovertible evidence from palaeoclimate records for a link between varying cosmic radiation and climate. Using samples from a speleothem from a cave in Oman, Middle East, these authors showed that a close correlation exists between radiocarbon production rates (driven by incoming cosmic radiation, which is solar modulated) and rainfall (as reflected in the geochemical signature of oxygen isotopes).

As already discussed, the 1500-year Bond Cycle is probably of solar origin. Another climate rhythm of similar length occurs in glacial sediments deposited from about 90,000 to 15,000 years ago, especially in the North Atlantic region (Dansgaard et al. 1971; Rahmstorf 2003). Called a Dansgaard-Oeschger, or D-O, event, this cycle may also be a response to solar forcing.

The IPCC represents this large corpus of research into Sun–Earth climate relations (and other non-greenhouse gas causes of climate change, such as 1800- and 5000-year lunar tidal cycles) (Keeling & Whorf 2000) poorly in its assessment reports, and nor has any significant attempt been made to include such solar effects in the IPCC's modelling. As a result, the current generation of general circulation models (GCM) under-estimate the variability and changes from solar-induced climate signals. That many of the mechanisms, and possible mechanisms, by which the Sun influences Earth's climate are poorly understood is no justification for ignoring them.

The average length of solar cycles, from minimum to minimum, is eleven years. The solar minimum between cycles 23 and 24 occurred in December 2008 (Marshall Space Flight Center 2010), making cycle 23, at 12.5 years long, the longest since 1823. However, the Sun remained in a quiescent state through most of 2009, with only inter-mittent cycle 24 sunspots occurring; by the end of 2009 there had been 771 days without sunspots during the transition.

It is established from observation that solar cycles longer than the eleven-year average are followed by later cycles of lesser intensity, and, commensurately, a cooler climate (Friis-Chistensen & Lassen 1991; Archibald 2006; Archibald 2007; Butler & Johnston 1996; Clilverd et al. 2006). Solar cycle 23 was three years longer than cycle 22. Based on the theory originally proposed by Friis-Christensen and Lassen, this implies that cooling of up to 2.2 °C may occur during cycle 24 (compared with temperatures during cycle 23) for the mid-latitude grain-growing areas of the Northern Hemisphere (Archibald 2010).

The Average Planetary Magnetic Index (AP index) is a commonly used yardstick to indicate the Earth's magnetic condition and its close relation to solar magnetic variability. Historically, this index recorded a lowest monthly value of 4 in January 1932. That is until recently, when, between November 2008 and September 2009, the AP index returned persistent readings of 4s and 5s, sinking to 3s in October and November, and finally to 1 in December. This is the lowest reading in the 165 years of observations, since 1844 (Watts 2010). At the same time that the Sun's magnetic field was setting new record lows, the Northern Hemisphere's winter turned into one of the most severe in decades. It was therefore not surprising that during 2009 NASA's chief solar scientist, Dr David Hathaway, commented that, 'something like the Dalton Minimum – two solar cycles in the early 1800s that peaked at about an average of 50 sunspots – lies in the realm of the possible' (Archibald 2009). Given the implications of the recurrence of the cold conditions that characterised the Dalton Minimum, it is astonishing that the IPCC, and the governments that it advises, continue to ignore the implications of the cycle 23–24 solar transition, and the now probable outcome of significant cooling and decrease of crop yields over the next few decades.

Most of the assertions that solar causes can only play a minor role in controlling the late twentieth-century warming are misplaced, because their authors not only fail to consider all the mechanisms listed above, but also fail to allow for the lags that occur in the redistribution of solar heat through various planetary heat transfers and storages. As US astrophysicist, Willie Soon has shown delays of five to twenty years in the manifestation of a solar signal are associated with ocean storage and circulation of heat (2009). The peak solar radiation outputs observed in the 1980s and early 1990s, with weakening since then, match well with climatic observations when applied with an appropriate time lag.

In summary, the argument advanced by the IPCC – that the Sun can only affect climate through irradiance, that irradiance changes are small, and so the Sun cannot play a major role in global warming or cooling trends – is incorrect.

The human influence on climate change, in context

Despite the great variability and high magnitudes of natural climate change, it is clearly also the case that human activities have a measurable effect on local climates.

For example, the concrete, glass, steel and macadam that are used to build a conurbation absorb more radiant heat from the Sun during the day than did the pre-existing natural vegetation. The result is a local warming called the urban heat island effect, which, for a large city, has a magnitude of several degrees (Hingane 1996; McKitrick & Michaels 2004; McKitrick & Michaels 2007; Brandsma et al. 2003).

Alternatively, when humans clear forested areas, the pasture or crops that are planted are often lighter in colour than was the forest. This results in reflection of more of the incoming solar energy than before, and hence cooling (Steyaert & Knox 2008; Fall et al. 2009; Pielke et al. 2007).

So humans, through changed land usage, have an effect on local climate that is variously warming or cooling. Summing these local signals all over the globe, it follows that humans must exercise an effect on global climate, also. The question in context, therefore, is not 'do humans have an effect on global climate?' But rather, 'what is the sign and magnitude of the net human effect on global climate?' As you will see, no one knows the answer.

Local, regional, and global climates change naturally all the time, and human activity is definitely known to cause local change. Yet, remarkably, given the expenditure and effort spent looking for it since 1990, no summed human effect on global climate has ever been identified or measured. Therefore, the human signal most probably lies buried in the variability and noise of the natural climate system. This is so to such a degree that as a statement of fact we cannot even be certain whether the net human signal is one of global warming or cooling (Carter et al. 2007; Pielke 2002; Cooling: The Human Climate Signal? 2008). Though it is true that many scientists anticipate on theoretical grounds that net warming is the more likely, no direct evidence exists that any such warming would, *ipso facto*, be dangerous.

22 Mass Death Dies Hard
Clive James

When you tell people once too often that the missing extra heat is hiding in the ocean, they will switch over to watch *Game of Thrones*, where the dialogue is less ridiculous and all the threats come true. The proponents of man-made climate catastrophe asked us for so many leaps of faith that they were bound to run out of credibility in the end.

Now that they finally seem to be doing so, it could be a good time for those of us who have never been convinced by all those urgent warnings to start warning each other that we might be making a comparably senseless tactical error if we expect the elastic cause of the catastrophists, and all of its exponents, to go away in a hurry.

I speak as one who knows nothing about the mathematics involved in modelling non-linear systems. But I do know quite a lot about the mass media, and far too much about the abuse of language. So I feel qualified to advise against any triumphalist urge to compare the apparently imminent disintegration of the alarmist cause to the collapse of a house of cards. Devotees of that fond idea haven't thought hard enough about their metaphor. A house of cards collapses only with a sigh, and when it has finished collapsing all the cards are still there.

Although the alarmists might finally have to face that they will not get much more of what they want on a policy level, they will surely, on the level of their own employment, go on wanting their salaries and prestige. To take a conspicuous if ludicrous case, the Australian climate star Tim Flannery will probably not, of his own free will, shrink back to the position conferred by his original metier, as an expert on the extinction of the giant wombat. He is far more likely to go on being, and wishing to be, one of the mass media's mobile oracles about climate. While that possibility continues, it will go on being dangerous to stand between him and a TV camera. If the giant wombat could have moved at that speed, it would still be with us.

The mere fact that few of Flannery's predictions have ever come even remotely true need not be enough to discredit him. The same fact, in the case of America's Professor Ehrlich, has left him untouched ever since he predicted that the world would soon run out of copper. In those days, when our current phase of the long discussion about man's attack on nature was just beginning, he predicted mass death by extreme cold. Lately he predicts mass death by extreme heat. But he has always predicted mass death by extreme something, and he is always Professor Ehrlich.

Actually, a more illustrative starting point for the theme of the permanently imminent climatic apocalypse might be taken as 3 August 1971, when the *Sydney Morning Herald* announced that the Great Barrier Reef would be dead in six months. After six months the reef had not died, but it has been going to die almost as soon as that ever since; making it a strangely durable emblem for all those who have wedded themselves to the notion of climate catastrophe.

The most exalted of all the world's predictors of reef death, President Obama, has still not seen the reef even now but he promises to go there one day when it is well again. Assurances that it has never really been sick won't be coming from his senior science adviser John Holdren. In the middle of 2016 some of the long-term experts on reef death began admitting that they had all been overdoing the propaganda.

After almost half a century of reef death prediction, this was the first instance of one group of reef death predictors telling another group to dial down the alarmism, or they would queer the pitch for everybody. But an old hand like Holdren knows better than to listen to sudden outbursts of moderation. Back in the day, when extreme cooling was the fashion, he was an extreme coolist. Lately he is an extreme warmist. He will surely continue to be an extremist of some kind, even if he has to be an extreme moderate. And after all, his boss was right about the ocean. In his acceptance speech at the 2008 Democratic convention, Obama said – and I truly wish that this were an inaccurate paraphrase – that people should vote for him if they wanted to stop the ocean rising. He got elected, and it didn't rise.

The notion of a count-down or a tipping point is very dear to both wings of this deaf shouting match, and really is of small use to either. On the catastrophist wing, whose 'narrative', as they might put it, would so often seem to be a synthesised film script left over from the era of surround-sound disaster movies, there is always a count-down to the tipping point. When the scientists are the main contributors to the script, the tipping point will be something like the forever forthcoming moment when the Gulf Stream turns upside down or the Antarctic ice sheet comes off its hinges, or any other extreme event which, although it persists in not happening, could happen sooner than we think (science correspondents who can write a phrase like 'sooner than we think' seldom realise that they might have already lost you with the word 'could').

When the politicians join in the writing, the dramatic language declines to the infantile. There are only 50 days (Gordon Brown) or 100 months (Prince Charles wearing his political hat) left for mankind to 'do something' about 'the greatest moral challenge … of our generation'. (Kevin Rudd, before he arrived at the Copenhagen climate shindig in 2009.)

When he left Copenhagen, Rudd scarcely mentioned the greatest moral challenge again. Perhaps he had deduced, from the confusion prevailing throughout the conference, that the chances of the world ever uniting its efforts to 'do something' were very small. Whatever his motives for backing out of the climate chorus, his subsequent career was an early demonstration that to cease being a chorister would be no easy retreat, because it would be a clear indication that everything you had said on the subject up to then had been said in either bad faith or ignorance. It would not be enough merely to fall silent. You would have to travel back in time, run for office in the Czech Republic instead of Australia, and call yourself Vaclav Klaus.

Australia, unlike Kevin Rudd, has a globally popular role in the climate movie because it looks the part. Common reason might tell you that a country whose contribution to the world's emissions is only 1.4% can do very little about the biggest moral challenge even if it manages to reduce that contribution to zero; but your eyes tell you that Australia is burning up. On the classic alarmist principle of 'just stick your head out of the window and look around you', Australia always looks like Overwhelming Evidence that the alarmists must be right. Even now that the global warming scare has completed its transformation into the climate change scare so that any kind of event at either end of the scale of temperature can qualify as a crisis, Australia remains the top area of interest, still up there ahead of even the melting North Pole, despite the Arctic's miraculous capacity to go on producing ice in defiance of all instructions from Al Gore. A 'C'-student to his marrow, and thus never quick to pick up any reading matter at all, Gore has evidently never seen the *Life* magazine photographs of America's nuclear submarine *Skate* surfacing through the North Pole in 1959. The ice up there is often thin, and sometimes vanishes. But it comes back, especially when someone sufficiently illustrious confidently predicts that it will go away for good.

After 4.5 billion years of changing, the climate that made outback Australia ready for Baz Luhrmann's view-finder looked all set to end the world tomorrow. History has already forgotten that the schedule for one of the big drought sequences in his movie *Australia* was wrecked by rain, and certainly history will never be reminded by the mass media, which loves a picture that fits the story. In this way, the polar bear balancing on the photo-shopped shrinking ice-floe will always have a future in show business, and the cooling towers spilling steam will always be up there in the background of the TV picture while the panel of experts discuss what Julia Gillard still calls 'carbon', her word for carbon dioxide. Pictures of her house near the beach in Adelaide, on the other hand, will never be used to illustrate satirical articles about a retired prophet of the rising ocean who buys a house near the beach, because there won't be any such articles. The full 97% of all satirists who dealt themselves out of the climate subject back at the start look like staying out of it until the end, even if they get satirised in their turn. One could blame them for their pusillanimity, but it would be useless, and perhaps unfair. Nobody will be able plausibly to call Emma Thompson dumb for spreading gloom and doom about the climate: she's too clever and too creative. And anyway, she might be right. Cases like Leonardo di Caprio and Cate Blanchett are rare enough to be called brave. Otherwise, the consensus of silence from the wits and thespians continues to be impressive. If they did wish to speak up for scepticism, however, they wouldn't find it easy when the people who run the big TV outlets forbid the wrong kind of humour. On *Saturday Night Live* back there in 2007, Will Ferrell, brilliantly pretending to be George W. Bush, was allowed to get every word of the global warming message wrong, but he wasn't allowed to disbelieve it.

Just as all branches of the modern media love a picture of something that might be part of the Overwhelming Evidence for climate change even if it is really a picture of something else, they all love a clock ticking down to zero, and if the clock never quite gets there then the motif can be exploited forever. But the editors and producers must face the drawback of such perpetual excitement: it gets perpetually less exciting. Numbness sets in, and there is time to think after all. Some of the customers might even start asking where this language of rubber numbers has been heard before.

It was heard from Swift. In *Gulliver's Travels* he populated his flying island of Laputa with scientists busily using rubber numbers to predict dire events. He called these scientists 'projectors'. At the basis of all the predictions of the projectors was the prediction that the Earth was in danger from a Great Comet whose tail was 'ten hundred thousand and fourteen' miles long. I should concede at this point that a sardonic parody is not necessarily pertinent just because it is funny; and that although it might be unlikely that the Earth will soon be threatened by man-made climate change, it might be less unlikely that the Earth will be threatened eventually by an asteroid, or let it be a Great Comet; after all, the Earth has been hit before.

That being said, however, we can note that Swift has got the language of artificial crisis exactly right, to the point that we might have trouble deciding whether he invented it, or merely copied it from scientific voices surrounding him in his day. James Hansen is a Swiftian figure. Blithely equating trains full of coal to trains full of people on their way to Auschwitz, Hansen is utterly unaware that he has not only turned the stomachs of the informed audience he was out to impress, he has lost their attention. Professor of Earth Sciences Chris Turney, who led a ship full of climate change enthusiasts into the Antarctic ice to see how the ice was doing under the influence of climate change and found it was doing well enough to trap the ship, could have been invented by Swift. (Turney's subsequent *Guardian* article, in which he explained how this embarrassment was due only to a quirk of the weather, and had nothing to do with a possible mistake about the climate, was a Swiftian lampoon in all respects.) Compulsorily retired now from the climate scene, Dr Rajendra Pachauri was a zany straight from Swift, by way of a Bollywood remake of *The Party* starring the local imitator of Peter Sellers; if Dr Johnson could have thought of Pachauri, *Rasselas* would be much more entertaining than it is. Finally, and supremely, Tim Flannery could have been invented by Swift after ten cups of coffee too many with Stella. He wanted to keep her laughing. Swift projected the projectors who now surround us.

They came out of the grant-hungry fringe of semi-science to infect the heart of the mass media, where a whole generation of commentators taught each other to speak and write a hyperbolic doom-language ('unprecedented', 'irreversible', etcetera), which you might have thought was sure to doom them in their turn. After all, nobody with an intact pair of ears really listens for long to anyone who talks about 'the planet' or 'carbon' or 'climate denial' or 'the science'. But for now – and it could be a long now – the advocates of drastic action are still armed with a theory that no fact doesn't fit. The theory has always been manifestly unfalsifiable, but there are few science pundits in the mass media who could tell Karl Popper from Mary Poppins. More startling than their ignorance, however, is their defiance of logic. You can just about see how a bunch of grant-dependent climate scientists might go on saying that there was never a Medieval Warm Period even after it has been pointed out to them that any old corpse dug up from the permafrost could never have been buried in it. But how can a bunch of supposedly enlightened writers go on saying that? Their answer, if pressed, is usually to say that the question is too elementary to be considered.

Alarmists have always profited from their insistence that climate change is such a complex issue that no 'science denier' can have an opinion about it worth hearing. For most areas of science such an insistence would be true. But this particular area has a knack of raising questions that get more and more complicated in the absence of an answer to the elementary ones. One of those elementary questions is about how man-made carbon dioxide can be a driver of climate change if the global temperature has not gone up by much over the last twenty years but the amount of man-made carbon dioxide has. If we go on to ask a supplementary question – say, how could carbon dioxide raise temperature when the evidence of the ice cores indicates that temperature has always raised carbon dioxide – we will be given complicated answers, but we still haven't had an answer to the first question, except for the suggestion that the temperature, despite the observations, really has gone up, but that the extra heat is hiding in the ocean. It is not necessarily science denial to propose that this long professional habit of postponing an answer to the first and most elementary question is bizarre. Richard Feynman said that if a fact doesn't fit the theory, the theory has to go. Feynman was a scientist. Einstein realised that the Michelson-Morley experiments hinted at a possible fact that might not fit Newton's theory of celestial mechanics. Einstein was a scientist too. Those of us who are not scientists, but who are sceptical about the validity of this whole issue – who suspect that the alleged problem might be less of a problem than is made out – have plenty of great scientific names to point to for exemplars, and it could even be said that we could point to the whole of science itself. Being resistant to the force of its own inertia is one of the things that science does.

When the climatologists upgraded their frame of certainty from global warming to climate change, the bet-hedging manoeuvre was so blatant that some of the sceptics started predicting in their turn; the alarmist cause must surely now collapse, like a house of cards. A tipping point had been reached. Unfortunately for the cause of rational critical enquiry, the campaign for immediate action against climate doom reaches a tipping point every few minutes, because the observations, if not the calculations, never cease exposing it as a fantasy. I myself, after I observed Andrew Neil on BBC TV wiping the floor with the then Secretary for Energy and Climate Change Ed Davey, thought that the British government's energy policy could not survive, and that the mad work which had begun with Ed Miliband's Climate Act of 2008 must now surely begin to come undone. Neil's well-informed list of questions had been a tipping point. But it changed nothing in the short term. It didn't even change the BBC, which continued uninterrupted with its determination that the alarmist view should not be questioned.

How did the upmarket mass media get themselves into such a condition of servility? One is reminded of that fine old historian George Grote, when he said that he had taken his *A History of Greece* only to the point where the Greeks themselves failed to realise they were slaves. The BBC's monotonous plugging of the climate theme in its science documentaries is too obvious to need remarking, but it's what the science programmes never say that really does the damage. Even the news programmes get 'smoothed' to ensure that nothing interferes with the constant business of protecting the climate change theme's dogmatic status. To take a simple but telling example: when Sigmar Gabriel, Germany's Vice Chancellor and man in charge of the *Energiewende*, talked rings around Greenpeace hecklers with nothing on their minds but renouncing coal, or told executives of the renewable energy companies that they could no longer take unlimited subsides for granted, these instructive moments could be seen on German television but were not excerpted and subtitled for British television even briefly, despite Gabriel's accomplishments as a natural TV star, and despite the fact that he himself was no sceptic.

Wrong message: easier to leave him out. And if the climate scientist Judith Curry appears before a US Senate committee and manages to defend her anti-alarmist position against concentrated harassment from a senator whose only qualification for the discussion is that he can impugn her integrity with a rhetorical contempt of which she is too polite to be capable? Leave it to YouTube. In this way, the BBC has spent ten years unplugged from a vital part of the global intellectual discussion, with an increasing air of provincialism as the inevitable result. As the UK now begins the long process of exiting the European Union, we can reflect that the departing nation's most important broadcasting institution has been behaving, for several years, as if its true aim were to reproduce the thought control that prevailed in the Soviet Union.

As for the print media, it's no mystery why the upmarket newspapers do an even more thorough job than the downmarket newspapers of suppressing any dissenting opinion on the climate. In Britain, the *Telegraph* sensibly gives a column to the diligently sceptical Christopher Booker, and Matt Ridley has recently been able to get a few rational articles into the *Times*, but a more usual arrangement is exemplified by my own newspaper, the *Guardian*, which entrusts all aspects of the subject to George Monbiot, who once informed his green readership that there was only one reason I could presume to disagree with him, and them: I was an old man, soon to be dead, and thus with no concern for the future of 'the planet'. I would have damned his impertinence, but it would have been like getting annoyed with a wheelbarrow full of freshly cut grass.

These byline names are stars committed to their opinion, but what's missing from the posh press is the non-star name committed to the job of building a fact-file and extracting a reasoned article from it. Further down the market, when the *Daily Mail* put its no-frills news-hound David Rose on the case after Climategate, his admirable competence immediately got him labelled as a 'climate change denier': one of the first people to be awarded that badge of honour. The other tactic used to discredit him was the standard one of calling his paper a disreputable publication. It might be – having been a victim of its prurience myself, I have no inclination to revere it – but it hasn't forgotten what objective reporting is supposed to be. Most of the British papers have, and the reason is no mystery.

They can't afford to remember. The print media are on their way down the drain. With almost no personnel left to do the writing, the urge at editorial level is to give all the science stuff to one bloke. The print edition of *The Independent* bored its way out of business when their resident climate nag was allowed to write half the paper. In its last year, when the doomwatch journalists were threatened by the climate industry with a newly revised consensus opinion that a mere two-degree increase in world temperature might be not only acceptable but likely, the *Independent's* chap retaliated by writing stories about how the real likelihood was an increase of five degrees, and in a kind of frenzied crescendo he wrote a whole front page saying that the global temperature was 'on track' for an increase of six degrees. Not long after, the *Indy's* print edition closed down.

At the *New York Times*, Andrew Revkin, star colour-piece writer on the climate beat, makes the whole subject no less predictable than his prose style: a cruel restriction. In Australia, the Fairfax papers, which by now have almost as few writers as readers, reprint Revkin's summaries as if they were the voice of authority, and will probably go on doing so until the waters close overhead. On the ABC, the house science pundit Robyn Williams famously predicted that the rising of the waters 'could' amount to 100 metres in the next century. But not even he predicted that it could happen next week. At the *Sydney Morning Herald*, it could happen next week. The only remaining journalists could look out of the window, and see fish.

Bending their efforts to sensationalise the news on a scale previously unknown even in their scrappy history, the mass media have helped to consolidate a pernicious myth. But they could not have done this so thoroughly without the accident that they are the main source of information and opinion for people in the academic world and in the scientific institutions. Few of those people have been reading the sceptical blogs: they have no time. If I myself had not been so ill during the relevant time-span, I might not have been reading them either, and might have remained confined within the misinformation system where any assertion of forthcoming disaster counts as evidence. The effect of this mountainous accumulation of sanctified alarmism on the academic world is another subject. Some of the universities deserve to be closed down, but I expect they will muddle through, if only because the liberal spirit, when it regains its strength, is likely to be less vengeful than the dogmatists were when they ruled. Finding that the power of inertia blesses their security as once it blessed their influence, the enthusiasts might have the sense to throttle back on their certitude, huddle under the blanket cover provided by the concept of 'post-normal science', and wait in comfort to be forgotten.

As for the learned societies and professional institutions, it was never a puzzle that so many of them became instruments of obfuscation instead of enlightenment. Totalitarianism takes over a state at the moment when the ruling party is taken over by its secretariat; the tipping point is when Stalin, with his lists of names, offers to stay late after the meeting and take care of business. The same vulnerability applies to any learned institution. Rule by bureaucracy favours mediocrity, and in no time at all you are in a world where Julia Slingo is a figure of authority, and Judith Curry is fighting to breathe. Under Stalin, Trofim Lysenko became more indispensable the more he reduced all the other biologists to the same condition as Soviet agriculture, and even after Stalin was dead, it took Andrei Sakharov to persuade Khrushchev not to bring Lysenko back to office. Khrushchev was well aware that Lysenko was a charlatan, but he looked like an historic force; and who argues with one of those?

On a smaller scale of influential prestige, Lord Stern lends the Royal Society the honour of his presence. For those of us who regard him as a vocalised stuffed shirt, it is no use saying that his confident pronouncements about the future are only those of an economist. Vaclav Klaus was only an economist when he tried to remind us that Malthusian clairvoyance is invariably a harbinger of totalitarianism. But Klaus was a true figure of authority. Alas, true figures of authority are in short supply, and tend not to have much influence when they get to speak.

All too often, this is because they care more about science than about the media. As recently as 2015, after a full ten years of nightly proof that this particular scientific dispute was a media event before it was anything, Freeman Dyson was persuaded to go on television. He was up there just long enough to say that the small proportion of carbon dioxide that was man-made could only add to the world's supply of plant food. The world's mass media outlets ignored the footage, mainly because they didn't know who he was. I might not have known either if I hadn't spent, in these last few years, enough time in hospitals to have it proved to me on a personal basis that real science is as indispensable for modern medicine as cheap power. Among his many achievements, to none of which he has ever cared about drawing attention, Dyson designed the TRIGA reactor. The TRIGA ensures that the world's hospitals get a reliable supply of isotopes.

Dyson served science. Except for the few hold-outs who go on fighting to defend the objective nature of truth, most of the climate scientists who get famous are serving themselves. There was a time when the journalists could have pointed out the difference, but now they have no idea. Instead, they are so celebrity-conscious that they would supply Tim Flannery with a new clown-suit if he wore out the one he is wearing now. In 2016, he dived on the Great Barrier Reef and reported himself overwhelmed by the evidence that it was on the point of death, a symptomatology which, he said, he had recently learned to recognise by watching his father die. Neither he nor any of his admirers at the *Sydney Morning Herald* cared to note that it has now been almost 50 years that the reef has been going to die soon. But the moment never came, although it will probably go on being about to happen for the next 50 years as well. The reef death disaster is like those millions of climate change refugees who were going to flood into the West by 2010. They never arrived. But when the refugees from the war in Syria started to arrive, there was a ready-made media apparatus waiting to declare that they were the missing climate change refugees really, because what else had caused the war but climate change? They were the missing heat that had been hiding in the ocean.

A bad era for science has been a worse one for the mass media, the field in which, despite the usual blunders and misjudgements, I was once proud to earn my living. But I have spent too much time, in these last few years, being ashamed of my profession: hence the note of anger which, I can now see, has crept into this essay even though I was determined to keep it out. As my retirement changed to illness and then to dotage, I would have preferred to sit back and write poems than to be known for taking a position in what is, despite the colossal scale of its foolish waste, a very petty quarrel. But when some of the climate priesthood, and even the Attorney General of the United States, started talking about how dissent might be suppressed with the force of law – well, that was a tipping point. I am a dissenter, and not because I deny science, but because I affirm it. So it was time to stand up and fight, if only because so many of the advocates, though they must know by now that they are professing a belief they no longer hold, will continue to profess it anyway.

Back in the day, when I was starting off in journalism – on the *Sydney Morning Herald*, as it happens – the one thing we all learned early from our veteran colleagues was never to improve the truth for the sake of the story. If they caught us doing so, it was the end of the world.

But here we are, and the world hasn't ended after all. Though some governments might not yet have fully returned to the principle of evidence-based policy, most of them have learned to be wary of policy-based evidence. They have learned to spot it coming, not because the real virtues of critical enquiry have been well argued by scientists, but because the false claims of abracadabra have been asserted too often by people who, though they might have started out as scientists of a kind, have found their true purpose in life as ideologists. Modern history since

World War II has shown us that it is unwise to predict what will happen to ideologists after their citadel of power has been brought low. It was feared that the remaining Nazis would fight on, as Werewolves. Actually, only a few days had to pass before there were no Nazis to be found anywhere except in Argentina, boring one another to death at the world's worst dinner parties.

After the collapse of the Soviet Union, on the other hand, when it was thought that no apologists for Marxist collectivism could possibly keep their credibility in the universities of the West, they not only failed to lose heart, they gained strength. Some critics would say that the climate change fad itself is an offshoot of this lingering revolutionary animus against liberal democracy, and that the true purpose of the climatologists is to bring about a world government that will ensure what no less a philanthropist than Robert Mugabe calls 'climate justice', in which capitalism is replaced by something more altruistic.

I myself prefer to blame mankind's inherent capacity for raising opportunism to a principle: the enabling condition for fascism in all its varieties, and often an imperative mind-set among high end frauds. On behalf of the UN, Maurice Strong, the first man to raise big money for climate justice, found slightly under a million dollars of it sticking to his fingers, and hid out in China for the rest of his life – a clear sign of his guilty knowledge that he had pinched it. Later operators lack even the guilt. They just collect the money, like the Prime Minister of Tuvalu, who has probably guessed by now that the sea isn't going to rise by so much as an inch; but he still wants, for his supposedly threatened atoll, a share of the free cash, and especially because the question has changed. It used to be: how will we cope when the disaster comes? The question now is: how will we cope if it does not?

There is no need to entertain visions of a vast, old-style army of disoccupied experts retreating through the snow, eating first their horses and finally each other. But there could be quite a lot of previously well-subsidised people left standing around while they vaguely wonder why nobody is listening to them anymore. Way back there in 2011, one of the Climategate scientists, Tommy Wils, with an engagingly honest caution rare among prophets, speculated in an email about what people outside their network might do to them if climate change turned out to be a bunch of natural variations: 'Kill us, probably.' But there has been too much talk of mass death already, and anyway most of the alarmists are the kind of people for whom it is a sufficiently fatal punishment simply to be ignored.

Nowadays I write with aching slowness, and by the time I had finished assembling the previous paragraph, the US had changed presidents. What difference this transition will make to the speed with which the climate change meme collapses is yet to be seen, but my own guess is that it was already almost gone anyway: a comforting view to take if you don't like the idea of a posturing zany like Donald Trump changing the world.

Personally, I don't even like the idea of Trump changing a light bulb, but we ought to remember that this dimwitted period in the history of the West began with exactly that: a change of light bulbs. Suddenly, 100 watts were too much. For as long as the climate change fad lasted, it always depended on poppycock; and it would surely be unwise to believe that mankind's capacity to believe in fashionable nonsense can be cured by the disproportionately high cost of a temporary embarrassment. I'm almost sorry that I won't be here for the ceremonial unveiling of the next threat. Almost certainly the opening feast will take place in Paris, with a happy sample of all the world's young scientists facing the fragrant remains of their first ever plate of foie gras, while vowing that it will not be the last.

References

1: The Extraordinary Resilience of Great Barrier Reef Corals, and Problems with Policy Science

Browman, HI 2016, 'Applying organised scepticism to ocean acidification research', *ICES Journal of Marine Science*, vol. 73, no. 3, pp. 529–536.

Brodie, J, De'ath, G, Devlin, M, Furnas, M & Wright, M 2007, 'Spatial and temporal patterns of near-surface chlorophyll *a* in the Great Barrier Reef lagoon', *Marine and Freshwater Research*, vol. 58, pp. 342–353.

De'ath, G, Lough, JM & Fabricius, KE 2009, 'Declining coral calcification on the Great Barrier Reef', *Science*, vol. 323, pp. 116–119.

De'ath, G, Fabricius, KE & Lough, JM 2013, 'Yes – Coral calcification rates have decreased in the last twenty-five years!' *Marine Geology*, vol. 346, pp. 400–402.

D'Olivo, JP, McCulloch, NT & Judd, K 2013, 'Long-term records of coral calcification across the central Great Barrier Reef: assessing the impacts of river runoff and climate change' *Coral Reefs*, vol. 32, pp. 999–1012.

Duarte, CM, Fulweiler, RW, Lovelock, CE, Martinetto, P, Saunders, MI, Pandolfi, JM, Stefan, G &Nixon, SW 2015, 'Reconsidering ocean calamities', *Bioscience*, vol. 65, no. 2, pp. 130–139.

Glasziou, P, Meats E, Heneghan, C & Shepperd, S 2008, 'What is missing from descriptions of treatment in trials and reviews?' *BMJ*, vol. 336, pp. 1472–1474.

Hartshorne, JK & Schachner, A 2012, 'Tracking replicability as a method of post-publication open evaluation', *Frontiers in Computational Neuroscience*, vol. 6, pp. 1–13.

Harrison, PL, Babcock, RC, Bull, GD, Oliver, JK, Wallace, CC & Willis, BL 1984, 'Mass spawning in tropical reef corals', *Science*, vol. 223, pp. 1186–1189.

Horton, R 2015, 'Offline: What is medicine's 5 sigma?' *Lancet*, vol. 385, p. 1380.

Hughes, TP 2016, 'Widespread coral bleaching detected on the Great Barrier Reef', ABC *Radio National Breakfast*, 2 March, viewed 16 February 2017, http://www.abc.net.au/radionational/programs/breakfast/widespread-coral-bleaching-detected-on-the/7212760

Ioannidis, JPA 2014, 'How to make more published research true', *PLoS Medicine*, vol. 11, no. 10.

Lloyd, G 2016, 'Reef whistle blower censured by James Cook University', *The Weekend Australian*, 11–12 June, p. 3.

Lough, JB & Barnes, DJ 2000, 'Environmental controls on growth of the massive coral *Porites.*', *Journal of Experimental Marine Biology and Ecology*, vol. 245, pp. 225–243.

Marshall, PA & Schuttenberg, HZ 2006, 'A Reef Manager's Guide to Coral Bleaching', Great Barrier Reef Marine Park Authority, Townsville, Australia.

Oliver, JK, Berkelmans, R & Eakin, CM 2009, 'Coral Bleaching in space and time' in MJH van Oppen & JM Lough (eds), *Coral Bleaching: Patterns, Processes, Causes and Consequences*, Springer Berlin Heidelberg.

Peason, RG & Endean, R 1969, 'A preliminary study of the coral predator *Acanthaster planci* (L.) (Asteroidea) on the Great Barrier Reef', Queensland Department of Harbour and Marine Fisheries, notes 3, pp. 27–55.

Prinz, F, Schlange, T & Asadullah, K 2011, 'Believe it or not: how much can we rely on published data on potential drug targets?' *Nature Reviews Drug Discovery*, vol. 10, p. 712.

Ridd, PV 2007, 'A critique of a method to determine long-term decline of coral reef ecosystems', *Energy and Environment*, vol. 18, no. 6, pp. 783–796.

Ridd, PV, Orpin, AR, Stieglitz, TC & Brunskill, GJ 2011, 'Will reducing agricultural runoff drive recovery of coral biodiversity and macroalgae cover on the Great Barrier Reef?' *Ecological Applications*, vol. 22, pp. 3332–3335.

Ridd, PV, Larcombe, P, Stieglitz, T & Orpin, A 2012, 'Comment on "Towards ecologically relevant targets for river pollutant loads to the Great Barrier Reef" by FJ Kroon', *Marine Pollution Bulletin*, vol. 65, no. 4–9, pp. 261–266.

Ridd, PV, DaSilva, ET & Stieglitz, TC 2013, 'Have coral calcification rates slowed in the last twenty years?' *Marine Geology*, vol. 346, pp. 392–399.

Roff, G, Bejarano, S, Bozec, Y, Nugues, M, Steneck, RS & Mumby, PJ 2014, '*Porites* and the Phoenix effect: unprecedented recovery after a mass coral bleaching event at Rangiroa Atoll, French Polynesia', *Marine Biology*, vol. 161, pp. 1385–1393.

Steele, J 2016, 'The Coral Bleaching Debate: Is Bleaching the Legacy of a Marvelous Adaptation Mechanism or A Prelude to Extirpation?' *Watts Up With That?* Viewed 17 February 2017, https://wattsupwiththat.com/2016/05/18/the-coral-bleaching-debate-is-bleaching-the-legacy-of-a-marvelous-adaptation-mechanism-or-a-prelude-to-extirpation/

Vasilevsky, NA, Brush, MH, Paddock, H, Ponting, L, Tripathy, SJ, LaRocca, GM & Haendel, MA 2013, 'On the reproducibility of science: unique identification of research resources in the biomedical literature', *PeerJ*, vol. 1.

Wagenmakers, EJ, Wetzels, R, Borsboom, D, van der Maas, HLJ & Kievit, RA 2012, 'An agenda for purely confirmatory research', *Perspectives on Psychological Science*, vol. 7, pp. 632–638.

Walbran, PW, Henderson, RA, Jull, AJT & Head, JM 1989, 'Evidence from sediments of long-term *Acanthaster planci* predation on corlas of the Great Barrier Reef', *Science*, vol. 245, pp. 847–850.

Yonge, CM 1930, Great Barrier Reef Expedition, 1928–29: Scientific Reports, British Museum.

2: Ocean Acidification: Not Yet a Catastrophe for the Great Barrier Reef

Anderson, KD, Heron, SF, Pratchett, MS 2015, 'Species-specific declines in the linear extension of branching corals at a subtropical reef, Lord Howe Island', *Coral Reefs*, vol. 34, pp. 479–490.

Banobi, JA, Branch, TA & Hilborn, R 2011, 'Do Rebuttals Affect Future Science?' *Ecosphere*, vol. 2, pp. 1–11.

Biello, D 2009, 'Ocean acidification hits Great Barrier Reef', *Scientific American*, viewed 17 February 2017, https://www.scientificamerican.com/article/ocean-acidification-hits-great-barrier-reef/

Brien, HV, Watson, S-A & Hoogenboom, MO 2016, 'Presence of competitors influences photosynthesis, but not growth, of the hard coral *Porites cylindrica* at elevated seawater CO_2', *ICES Journal of Marine Science*, vol. 73, pp. 659–669.

Browman, HI 2016, 'Applying organised scepticism to ocean acidification research', *ICES Journal of Marine Science*, vol. 73, no. 3, pp. 529–536.

Carter, RM 2010, *Climate: The Counter Consensus*, Stacey International, p. 315.

Chua, CM, Leggat, W, Moya, A & Baird, AH 2013a, 'Near-future reductions in pH have no consistent ecological effects on the early life history stages of reef corals', *Marine Ecology Progress Series*, vol. 486, pp. 143–151.

Chua, CM, Leggat, W, Moya, A & Baird, AH 2013b, 'Temperature affects the early life history stages of corals more than near future ocean acidification,' *Marine Ecology Progress Series*, vol. 475, pp. 85–92.

De'ath, G, Lough, JM & Fabricius, KE 2009, 'Declining coral calcification on the Great Barrier Reef', *Science*, vol. 323, pp. 116–119.

D'Olivo, JP, McCulloch, NT & Judd, K 2013, 'Long-term records of coral calcification across the central Great Barrier Reef: assessing the impacts of river runoff and climate change' *Coral Reefs*, vol. 32, pp. 999–1012.

Dore, JE, Lukas, R, Sadler, DW, Church, M & Karl, DM 2009, 'Physical and biogeochemical modulation of ocean acidification in the central North Pacific', *PNAS*, vol. 106, no. 30, pp. 12235–12240.

Eriander, L, Wrange, A-L & Havenhand, JN 2016, 'Simulated diurnal pH fluctuations radically increase *variance* in – but not the *mean* of – growth in the barnacle *Balanus improvisus*', *ICES Journal of Marine Science*, vol. 73, pp. 596–603.

Gattuso, JP & Hansson, L (eds) 2011, *Ocean Acidification*, Oxford University Press.

Gattuso, JP, Kirkwood, W, Barry, JP, Cox, E, Gazeau, F, Hansson, L, Hendriks, I, Kline, DI, Mahacek, P, Martin, S, McElhany, P, Peltzer, ET, Reeve, J, Roberts, D, Saderne, V, Tait, K, Widdicombe, S & Brewer, PG 2015, 'Free-ocean CO2 enrichment (FOCE) systems: present status and future developments', *Biogeosciences*, vol. 11, pp. 4057–4075.

Georgiou L, Falter J, Trotter J, Kline DI, Holcomb M, Dove SG, Hoegh-Guldberg O & McCulloch M 2015, 'pH homeostasis during coral calcification in a FOCE experiment, Heron Island reef flat, Great Barrier Reef', *PNAS*, vol. 112, no. 43, pp. 13219–13224.

Hofmann, GE, Smith, JE, Johnson, KS, Send, U, Levin, LA, Micheli, F, Paytan, A, Price, NN, Peterson, B Takeshita, Y, Matson, PG, Derse Crook, E, Kroeker, KJ, Gambi, MC, Rivest, EB, Frieder, CA, Yu, PC & Martz, TR 2011, 'High-frequency dynamics of ocean pH: a multi-ecosystem comparison', *PLoS ONE*, vol. 6, no. 12, e28983.

Kline, DI, Teneva, L, Hauri, C, Schneider, K, Miard, T, Chai, A, Marker, M, Dunbar, R, Caldeira, K, Lazar, B, Rivlin, T, Mitchell, BG, Dove, S & Hoegh-Guldberg, O 2015, 'Six month *in situ* high-resolution carbonate chemistry and temperature study on a coral reef flat reveals asynchronous pH and temperature anomalies,' *PLoS ONE*, vol. 10, no. 6, e0127648.

Lawson, N 2008, *An Appeal to Reason: A Cool Look at Global Warming*, Duckworth Overlook, London.

Lewis, N, Brown, KA, Edwards, LA, Cooper, G & Findlay, HS 2013, 'Sensitivity to ocean acidification parallels natural pCO_2 gradients experienced by Arctic copepods under winter sea ice', *PNAS*, vol. 110, pp. 4960–4967.

Lohbeck, KT, Riebesell, U & Reusch, R 2012, 'Adaptive evolution of a key phytoplankton species to ocean acidification', *Nature Geoscience*, vol. 5, no. 5, pp. 346–351.

Lohbeck, KT, Riebesell, U & Reusch, R 2014, 'Gene expression changes in the coccolithophore *Emiliania huxleyi* after 500 generations of selection to ocean acidification', *Proc Biol Sci*, vol. 281, no. 1786.

Liu, Y, Liu, W, Peng, Z, Xiao, Y, Wei, G, Sun, W, He, J, Liu, G & Chou, C-L 2008, 'Instability of seawater pH in the South China Sea during the mid-late Holocene: Evidence from boron isotopic composition of corals', *Geochimica et Cosmochimica Acta*, vol. 73, iss. 5, pp. 1264–1272.

Miller, GM, Watson, S-A, McCormick, MI & Munday, PL 2013, 'Increased CO_2 stimulates reproduction in a coral reef fish', *Global Change Biology*, vol. 19, no. 10, pp. 3037–3045.

Miller, GM, Watson, S-A, Donelson, JM, McCormick, MI & Munday, PL 2012, 'Parental environment mediates impacts of increased carbon dioxide on a coral reef fish', *Nature Climate Change*, vol. 2, no. 12, pp. 858–861.

Ow, YX, Collier, CJ & Uthicke, S 2015, 'Responses of three tropical seagrass species to CO_2 enrichment', *Marine Biology*, vol. 162, no. 5, pp. 1005–1017.

Pelejero, C, Calvo, E, McCulloch, MT, Marshall, JF, Gagan, MK, Lough, JM & Opdyke, BN 2005, 'Preindustrial to Modern Interdecadal Variability in Coral Reef pH' *Science*, vol. 309, no. 5744, pp. 2204–2207.

Raisman, S & Murphy, DT 2013, *Ocean Acidification: Elements and Considerations*, Nova Science Publishers.

Revelle, R & Fairbridge, R 1957, 'Carbonates and carbon dioxide', Geological Society of America, memoir 67, pp. 239–285.

Riebesell, U, & Gattuso, JP 2015, 'Commentary: lessons learned from ocean acidification research', *Nature Climate Change*, vol. 5, no. 1, pp. 12–14.

Ries, JB, Cohen, AL & McCorkle, DC 2009, 'Marine calcifiers exhibit mixed responses to CO_2-induced ocean acidification', *Geology*, vol. 37, no. 12, pp. 1131–1134.

Ridd, PV, DaSilva, ET & Stieglitz, TC 2013, 'Have coral calcification rates slowed in the last twenty years?' *Marine Geology*, vol. 346, pp. 392–399.

Waldbusser, GG & Salisbury, JE 2014, 'Ocean acidification in the coastal zone from an organism's perspective: multiple system parameters, frequency domains, and habitats', *Annual Reviews of Marine Science*, vol. 6, pp. 221–247.

3: Understanding Climate Change in terms of Natural Variability

Abreu, JA, Beer, J, Ferriz-Mas, A, McCracken, KG & Steinhilber, F 2012, 'Is there a planetary influence on solar activity?' *Astronomy and Astrophysics*, vol. 548, no. A88.

Alley, RB 2004, 'GISP2 Ice Core Temperature and Accumulation Data', IGBP PAGES/World Data Center for Paleoclimatology Data Contribution Series #2004-013, NOAA/NGDC Paleoclimatology Program, Boulder CO, USA.

Bond, G, Kromer, B, Beer, J, Muscheler, R, Evans, MN, Showers, W, Hoffman, S, Lotti-Bond, R, Hajdas, I & Bonani, G 2001, 'Persistent solar influence on North Atlantic climate during the Holocene', *Science*, vol. 294, no. 5549, pp. 2130-2136.

Christiansen, B & Ljungqvist, FC 2012, 'The extra-tropical Northern Hemisphere temperature in the last two millennia: reconstructions of low-frequency variability', *Climate of the Past*, vol. 8, pp. 765-786.

Crowley, TJ 2000, 'Causes of climate change over the past 1000 years', *Science*, vol. 289, no. 5477, pp. 270–277.

Gervais, F 2016, 'Anthropogenic CO2 warming challenged by 60-year cycle', *Earth-Science Reviews*, vol. 155, pp. 129-135.

Haigh, ID, Eliot, M & Pattiaratchi, C 2011, 'Global influences of the 18.61 year nodal cycle and 8.85 year cycle of lunar perigee on high tidal levels,' *Journal of Geophysical Research Oceans*, vol. 116 (2011), no. C6.

Hoyt, DV & Schatten, KH 1997, *The Role of the Sun in the Climate Change*, Oxford University Press, New York.

Hung, C-C 2007, 'Apparent relations between solar activity and solar tides caused by the planets', NASA Glenn Research Center, Cleveland.

IPCC 2013, 'Climate Change 2013: The Physical Science Basis. Contribution of Working Group I to the Fifth Assessment Report of the Intergovernmental Panel on Climate Change', in TF Stocker et al (eds.), Cambridge University Press, Cambridge, UK.

Kerr, RA 2001, 'A variable sun paces millennial climate', *Science*, vol. 294, no. 5546, pp. 1431-1433.

Kirkby, J 2007, 'Cosmic rays and climate', *Surveys in Geophysics*, vol. 28, pp. 333-375.

Lamb, HH 1965, 'The earlier medieval warm epoch and its sequel', *Palaeogeography, Palaeoclimatology, Palaeoecology*, vol. 1, pp. 13-37.

Lamb, HH 1982, *Climate, History and the Modern World*, Routledge, New York.

Lewis, N 2013, 'An objective Bayesian improved approach for applying optimal fingerprint techniques to estimate climate sensitivity', *Journal of Climate*, vol. 26, pp. 7414-7429.

Liu, Z, Zhu, J, Rosenthal, Y, Zhang, X, Otto-Bliesnere, BL, Timmermannf, A, Smith, RS, Lohmannd, G, Zhengh, W & Timmi, OS 2014, 'The Holocene temperature conundrum', *PNAS*, vol. 111, no. 34.

Ljungqvist, FC 2010, 'A new reconstruction of temperature variability in the extra-tropical Northern Hemisphere during the last two millennia', *Geografiska Annaler*, vol. 92, pp 339-351.

Marcott, SA, Shakun, JD, Clark, PU & Mix, AC 2013, 'A reconstruction of regional and global temperature for the past 11,300 years', *Science*, vol. 339, pp. 1198-1201.

Mann, ME, Bradley, RS & Hughes, MK 1999, 'Northern hemisphere temperatures during the past millennium: inferences, uncertainties, and limitations', *Geophysical Research Letters*, vol. 26, no. 6, pp. 759-762.

Mann, ME, Zhang, Z, Hughes, MK, Bradley, MK, Miller, SK, Rutherford, S & Ni, F 2008, 'Proxy-based reconstructions of hemispheric and global surface temperature variations over the past two millennia', *PNAS*, vol. 105, no. 36, pp. 13252-13257.

Manzi, V, Gennari, R, Lugli, S, Roveri, M, Scafetta, N & Schreiber, C 2012, 'High-frequency cyclicity in the Mediterranean Messinian evaporites: evidence for solar-lunar climate forcing', *Journal of Sedimentary Research*, vol. 82, pp. 991-1005.

McCracken, KG, Beer, J & Steinhilber, F 2014 'Evidence for planetary forcing of the cosmic ray intensity and solar activity throughout the past 9400 years', *Solar Physics*, vol. 289, no. 8, pp. 3207-3229.

Meehl, GA, Arblaster, JM, Fasullo, JT, Hu, A & Trenberth, KE 2011, 'Model-based evidence of deep-ocean heat uptake during surface-temperature hiatus periods', *Nature Climate Change*, vol. 1, pp. 360-364.

Moberg, A, Sonechkin, DM, Holmgren, K, Datsenko, NM & Karlen, W 2005, 'Highly variable Northern Hemisphere temperatures reconstructed from low and high resolution proxy data', *Nature*, vol. 433, pp. 613-617.

Morice, CP, Kennedy, JJ, Rayner, NA & Jones, PD 2012, 'Quantifying uncertainties in global and regional temperature change using an ensemble of observational estimates: The HadCRUT4 dataset', *Journal of Geophysical Research*, vol. 117, no. D8.

Morner, N-A 2013, 'Planetary beat and solar-terrestrial responses', *Pattern Recognition in Physics*, vol. 1, pp. 107-116.

Ollila, A 2015, 'Cosmic theories and greenhouse gases as explanations of global warming', *Journal of Earth Sciences and Geotechnical Engineering*, vol. 5, no. 4, pp. 27-43.

Scafetta, N 2010, 'Empirical evidence for a celestial origin of the climate oscillations and its implications', *Journal of Atmospheric and Solar-Terrestrial Physics*, vol. 72, pp. 951-970.

Scafetta, N 2012a, 'Testing an astronomically based decadal-scale empirical harmonic climate model versus the IPCC (2007) general circulation climate models', *Journal of Atmospheric and Solar-Terrestrial Physics*, vol. 80, pp. 124-137.

Scafetta, N 2012b, 'A shared frequency set between the historical mid-latitude aurora records and the global surface temperature', *Journal of Atmospheric and Solar-Terrestrial Physics*, vol. 74, pp. 145-163.

Scafetta, N 2012c, 'Does the Sun work as a nuclear fusion amplifier of planetary tidal forcing? A proposal for a physical mechanism based on the mass-luminosity relation', *Journal of Atmospheric and Solar-Terrestrial Physics*, vol. 81-82, pp. 27-40.

Scafetta, N 2012d, 'Multi-scale harmonic model for solar and climate cyclical variation throughout the Holocene based on Jupiter–Saturn tidal frequencies plus the 11-year solar dynamo cycle', *Journal of Atmospheric and Solar-Terrestrial Physics*, vol. 80, pp.296-311.

Scafetta, N 2013, 'Discussion on climate oscillations: CMIP5 general circulation models versus a semi-empirical harmonic model based on astronomical cycles', *Earth-Science Reviews*, vol. 126, pp. 321-357.

Scafetta, N 2014, 'Discussion on the spectral coherence between planetary, solar and climate oscillations: a reply to some critiques', *Astrophysics and Space Science*, vol. 354, pp. 275-299.

Scafetta, N 2016, 'High resolution coherence analysis between planetary and climate oscillations', *Advances in Space Research*, vol. 57, pp. 2121-2135.

Scafetta, N, Milani, F, Bianchini, A & Ortolani, S 2016, 'On the astronomical origin of the Hallstatt oscillation found in radiocarbon and climate records throughout the Holocene', *Earth-Science Reviews*, vol. 162, pp. 24-43.

Scafetta, N & Willson, RC 2013a, 'Planetary harmonics in the historical Hungarian aurora record (1523–1960)', *Planetary and Space Science*, vol. 78, pp. 38-44.

Scafetta, N & Willson, RC 2013b, 'Empirical evidences for a planetary modulation of total solar irradiance and the TSI signature of the 1.09-year Earth–Jupiter conjunction cycle', *Astrophysics and Space Science*, vol. 348, no. 1, pp. 25-39.

Shakun, J, Clark, PU, He, F, Marcott, SA, Mix, AC, Liu, Z, Otto-Bliesner, B, Schmittner, A & Bard, E 2012, 'Global warming preceded by increasing carbon dioxide concentrations during the last deglaciation', *Nature*, vol. 484, pp. 49-54.

Solheim, J-E 2013, 'Signals from the planets, via the Sun to the Earth', *Pattern Recognition in Physics*, vol. 1, pp. 177-184.

Steinhilber, F, Abreu, JA, Beer, J, Brunner, I, Christl, M, Fischer, H, Heikkilä, U, Kubik, PW, Mann, M, McCracken, KG, Miller, H, Miyahara, H, Oerter, H & Wilhelms, F 2012, '9,400 years of cosmic radiation and solar activity from ice cores and tree rings', *PNAS*, vol. 109, pp. 5967-5971.

Stocker, TF, Quin, D, Plattner, GK, Tignor, M, Allen, SK, Boschung, J, Nauels, A, Xia, Y, Bex, Midgely, PM (eds.) 2013, *Climate Change 2013: The Physical Science Basis, The Intergovernmental Panel on Climate Change (IPCC)*, Cambridge University Press, Cambridge, United Kingdom and New York, NY, USA.

Svensmark, H, Bondo, T & Svensmark, J 2009, 'Cosmic ray decreases affect atmospheric aerosols and clouds', *Geophysical Research Letters*, vol. 36.

Svensmark, J, Enghoff, MB & Svensmark, H 2012, 'Effects of cosmic ray decreases on cloud microphysics', *Atmospheric Chemistry and Physics*, Discuss. 12, pp. 3595-3617.

Wang, Z, Wu, D, Song, X, Chen, X & Nicholls, S 2012, 'Sun–Moon gravitation-induced wave characteristics and climate variation', *Journal of Geophysical Research*, vol. 117.

Wilson, IRG 2013, 'The Venus–Earth–Jupiter spin-orbit coupling model', *Pattern Recognition Physics*, vol. 1, pp. 147-158.

Wolf, R 1859, 'Extract of a letter to Mr. Carrington', *Monthly Notices of the Royal Astronomical Society*, vol. 19, pp. 85-86.

Wolff, CL & Patrone, PN 2011, 'A new way that planets can affect the Sun', *Solar Physics*, vol. 266, pp. 227-246.

Wyatt, MG & Curry, JA 2014, 'Role for Eurasian Arctic shelf sea ice in a secularly varying hemispheric climate signal during the 20th century', *Climate Dynamics*, vol. 42, pp. 2763-2782.

5: Creating a False Warming Signal in the US Temperature Record

City of Las Vegas, Office of Sustainability 2010, 'Summary Report, Urban Heat Island Effect', April, viewed 19 June, 2016, http://www.coolrooftoolkit.org/wp-content/uploads/2012/06/Las-Vegas_UHI_Report_2010-2.pdf.

DePalma, A 2008, 'Weather History Offers Insight Into Global Warming', *The New York Times*, September 15, viewed 17 June, 2016, http://www.nytimes.com/2008/09/16/science/earth/16moho.html?_r=0.

Goetz, J 2008, 'Adjusting Pristine Data', *Watts Up With That?*, September 23, viewed 16 June, 2016, https://wattsupwiththat.com/2008/09/23/adjusting-pristine-data/.

Hinkel, KM, Nelson, FE, Klene, AE & Bell, JH 2003, 'The Urban Heat Island in Winter at Barrow, Alaska', *International Journal of Climatology*, vol. 23, pp. 1889-1905, viewed 1 July, 2016, http://onlinelibrary.wiley.com/doi/10.1002/joc.971/pdf.

Leroy, M 2010, 'Siting Classification for Surface Observing Stations on Land, Climate, and Upper-air Observations', JMA/WMO Workshop on Quality Management in Surface, Tokyo, Japan, 27–30 July.

Murray, J & Heggie, D 2016, 'From urban to national heat island: The effect of anthropogenic heat output on climate change in high population industrial countries', *Earths' Future*, vol. 4, pp. 298-304, viewed 5 April, 2017, http://onlinelibrary.wiley.com/doi/10.1002/2016EF000352/epdf.

National Climatic Data Center 2007, 'Marysville Fire Department, B91 Form, Record of River and Climatological Observations', October 2007, viewed 16 June, 2016, https://wattsupwiththat.files.wordpress.com/2009/06/marysville-b91-10-2007.pdf.

National Oceanic and Atmospheric Administration 2012a, 'Time of Observation Bias Adjustments', October 4, viewed 17 June, 2016, http://www.ncdc.noaa.gov/oa/climate/research/ushcn/#biasadj.

National Oceanic and Atmospheric Administration 2012b, 'Homogeneity Testing and Adjustment Procedures', October 4, viewed 17 June, 2016, http://www.ncdc.noaa.gov/oa/climate/research/ushcn/#homogeneity.

National Oceanic and Atmospheric Administration 2012c, 'Estimation of Missing Values', October 4, viewed 17 June, 2016, http://www.ncdc.noaa.gov/oa/climate/research/ushcn/#missing.

National Weather Service, Office of Climate, Weather, and Water Services 2006, 'Instructions for WS Form B-91', November 21, viewed 16 June, 2016, http://www.nws.noaa.gov/om/coop/forms/b91-notes.htm.

National Weather Service, Office of Climate, Weather, and Water Services 2016, 'Highest and lowest temperatures ever recorded' in Las Vegas, January 14, viewed 19 June, 2016, http://www.wrh.noaa.gov/vef/climate/LasVegasClimateBook/Highest%20and%20Lowest%20Maximum%20Temperatures%20Ever%20Recorded.pdf.

'ORD – A casual history of O'Hare Field, Chicago' n.d., viewed 19 June, 2016. http://www.cirrusimage.com/ORD.htm.

Reiss, B 2010, 'Barrow, Alaska: Ground Zero for Climate Change', *Smithsonian Magazine*, March, viewed 2 July, 2016, http://www.smithsonianmag.com/science-nature/barrow-alaska-ground-zero-for-climate-change-7553696.

Shein, KA 2008, 'Interactive quality assurance practices. Preprints, 24th Conf. on Interactive Information Processing Systems (IIPS)', *American Meteorology Society*, viewed 16 June, 2016, https://ams.confex.com/ams/88Annual/techprogram/paper_131217.htm.

Stachelski, C & Gorelow, A 2016, *The Climate of Las Vegas*, Nevada, National Weather Service, viewed June 19, 2016, http://www.wrh.noaa.gov/vef/climate/LasVegasClimateBook/ClimateofLasVegas.pdf.

United States Census Bureau 2016, 'International Data Base World Population: 1950–2050', viewed 20 June, 2016, http://www.census.gov/population/international/data/idb/worldpopgraph.php.

Watts, A 2009, 'How not to measure temperature, part 86: when in Rome, don't do as the Romans do', *Watts Up With That?*, March 28, viewed 4 July, 2016, https://wattsupwiththat.com/2009/03/28/how-not-to-measure-temperature-part-86-when-in-rome-dont-do-as-the-romans-do/.

Watts, A 2015, Comparison of temperature trends using an unperturbed subset of The US Historical Climatological Network, AGU Fall Meeting, San Francisco, December 14-18.

Miller, B 2016, '2015 is warmest year on record, NOAA and NASA say', *CNN*, 20 January, viewed 3 July, 2016, http://www.cnn.com/2016/01/20/us/noaa-2015-warmest-year/.

6: It was Hot in the USA – in the 1930s

EPA 2016, 'Climate Change Indicators: High and Low Temperatures', viewed June 14, 2016, https://www3.epa.gov/climatechange/science/indicators/weather-climate/high-low-temps.html.

'49 former NASA scientists go ballistic over agency's bias over climate change' 2012, *Business Insider*, 11 April, viewed 3 March 2017, http://business.financialpost.com/business-insider/49-former-nasa-scientists-go-ballistic-over-agencys-bias-over-climate-change.

Hansen, J, Ruedy, R, Glascoe, J & Sato, M 1999, 'Wither US Climate?' National Aeronautics and Space Administration Goddard Institute for Space Studies, viewed 3 March 2017, http://www.giss.nasa.gov/research/briefs/hansen_07/.

NASA 2016, 'July 2016 was the hottest month on record', NASA, viewed 3 March 2017, http://earthobservatory.nasa.gov/IOTD/view.php?id=88607.

NASA 2012, 'NASA finds 2012 sustained long-term climate warming trend', NASA, viewed 3 March 2017, http://www.nasa.gov/topics/earth/features/2012-temps.html.

7: Taking Melbourne's Temperature

eMelbourne, the city past & present n.d., viewed 13 March 2017, http://www.emelbourne.net.au/biogs/EM00360b.htm.

Marohasy, J & Abbot, J 2015, 'Assessing the quality of eight different maximum temperature time series as inputs when using artificial neural networks to forecast monthly rainfall at Cape Otway, Australia', *Atmospheric Research*, vol. 166, pp. 141-149.

McAneney, K, Salinger, MJ, Porteous, AS & Barber, RF 1990, Modification of an orchard climate with increasing shelter-belt height. *Agricultural and Forest Meteorology*, vol. 49, pp. 177-189.

8: Mysterious Revisions to Australia's Long Hot History

Australian Government, Department of the Environment and Energy 2014, 'Portfolio Budget Statements 2014-15', viewed 4 April 2017, http://www.environment.gov.au/about-us/publications/budget/portfolio-budget-statements-2014-15.

Commonwealth of Australia 2015, 'Australian Climate Observations Reference Network–Surface Air Temperature (ACORN-SAT) – Report of the Technical Advisory Forum', viewed 4 April 2017, http://www.bom.gov.au/climate/change/acorn-sat/documents/2015_TAF_report.pdf.

Deacon, EL 1952, 'Climatic change in Australia since 1880', *Australian Journal of Physics*, vol. 6, pp. 209-218, http://www.publish.csiro.au/PH/pdf/PH530209.

Ewart, JN 1861, 'Mr Stuart's Party', *The Cornwall Chronicle*, 23 January, p. 5, viewed 3 April 2017, http://trove.nla.gov.au/newspaper/article/65567656.

Hughes, WS 1995, Comment on DE Parker, 'Historical Thermometer Exposures in Australia', *International Journal of Climatology*, vol. 17, pp. 197-199, viewed 3 April, 2017, http://www.warwickhughes.com/papers/ozstev.htm.

Lloyd, G 2014a, Bureau of Meteorology 'altering climate figures', *The Australian*, 23 August, viewed 3 April, 2017, http://www.theaustralian.com.au/national-affairs/climate/bureau-of-meteorology-altering-climate-figures/news-story/5bccf49433ae80f23b332d3493437106.

Lloyd, G 2014b, 'Weatherman's records detail heat that "didn't happen"', *The Australian*, 30 August, viewed 3 April, 2017, http://www.theaustralian.com.au/national-affairs/climate/weathermans-records-detail-heat-that-didnt-happen/news-story/80b19c3b5d85ba6ce31e24bbb1221b1f.

Marohasy, J 2017, 'Australia's hottest day on record ever – deleted', *Jennifer Marohasy*, viewed 3 April 2017, http://jennifermarohasy.com/2017/02/australias-hottest-day-record-ever-deleted/.

Marohasy, J & Abbot, J 2016, 'Southeast Australian Maximum Temperature Trends, 1887–2013', in D Easterbrook (ed.), *Evidence-based Climate Science*, 2nd edn, Elsevier.

Mitchell, TL 1846, *Journal of an Expedition into the Interior of Tropical Australia*, eBooks@Adelaide, The University of Adelaide, viewed 3 April 2017, https://ebooks.adelaide.edu.au/m/mitchell/thomas/tropical/complete.html.

Nicholls, N, Tapp, R, Burrows, K & Richards, D 1996, Historical Thermometer Exposures in Australia. *International Journal of Climatology*, vol. 16, pp. 705-710.

Nova, J 2010, 'Australian warming trend adjusted UP by 40%', *Jo Nova*, July, viewed 3 April 2017, http://joannenova.com.au/2010/07/australian-warming-trend-adjusted-up-by-40/.

Nova, J 2011, 'Announcing a formal request for the Auditor General to audit the Australian BOM', *Jo Nova*, February, viewed 3 April 2017, http://joannenova.com.au/2011/02/announcing-a-formal-request-for-the-auditor-general-to-audit-the-australian-bom/.

Nova, J 2014a, 'Was the Hottest Day Ever in Australia not in a desert, but in far south Albany?!', *Jo Nova*, July, viewed 3 April 2017, http://joannenova.com.au/2014/07/was-the-hottest-day-ever-in-australia-not-in-a-desert-but-in-far-south-albany/.

Nova, J 2014b, 'The lost climate knowledge of Deacon 1952: hot dry summers from 1880-1910', *Jo Nova*, September, viewed 3 April 2017, http://joannenova.com.au/2014/09/the-lost-climate-knowledge-of-deacon-1952-australian-summers-were-hotter-from-1880-1910/.

Nova, J 2014c, '1953 Headline: Melbourne's weather is changing! Summers getting colder and wetter', *Jo Nova*, September, viewed 3 April 2017, http://joannenova.com.au/2014/09/1953-headline-melbournes-weather-is-changing-summers-getting-colder-and-wetter/.

Nova, J 2014d Australian BOM "neutral" adjustments increase minima trends up 50%, viewed 27 May 2017, http://joannenova.com.au/2014/06/. australian-bom-neutral-adjustments-increase-minima-trends-up-60/

Nova, J 2015a, 'If it can't be replicated, it isn't science: BOM admits temperature adjustments are secret', *Jo Nova*, September, viewed 4 April 2017, http://joannenova.com.au/2015/06/if-it-cant-be-replicated-it-isnt-%20 science-bom-admits-temperature-adjustments-are-secret/.

Nova, J 2015b, 'Camouflage illusions in the matrix: same mysterious temperature, same day, year after year' viewed 26 May 2017, http://joannenova.com. au/2015/09/camouflage-illusions-in-the-matrix-same-mysterious-temperature-same-day-year-after-year/.

Stewart, K 2010, 'The Australian Temperature Record- Part 9: An Urban Myth', *Kenskingdom, A reality check on global warming*, September, viewed 3 April 2017, https://kenskingdom.wordpress.com/2010/09/14/the-australian-temperature-record-part-9-an-urban-myth/.

Stewart, K 2014a, 'The Australian Temperature Record Revisited: Part 4 – Outliers', *Kenskingdom, A reality check on global warming*, February, viewed 3 April 2017, https://kenskingdom.wordpress.com/2014/07/16/the-australian-temperature-record-revisited-part-4-outliers/.

Stewart, K 2014b The Australian Temperature Record Revisited Part 3: Remaining Sites, viewed 27 May 2017, https://kenskingdom.wordpress.com/2014/06/15/the-australian-temperature-record-revisited-part-3-remaining-sites/.

Sturt, C 1849, *Narrative of an expedition into Central Australia*, eBooks@Adelaide, The University of Adelaide, viewed 12 April 2017, Chapters 5, 6 and 12, https://ebooks.adelaide.edu.au/s/sturt/charles/s93n/contents.html.

Trewin, B 1997, 'Another look at Australia's record high temperature', *Australian Meteorological Magazine*, vol. 46, pp. 251-256.

Trewin, B 2012, *Techniques involved in developing the Australian Climate Observations Reference Network – Surface Air Temperature (ACORN-SAT) dataset'*, The Centre for Australian Weather and Climate Research, March, viewed 3 April, 2017, http://www.cawcr.gov.au/technical-reports/CTR_049.pdf.

Williams, G 1953, 'Melbourne's Weather is Changing', *The Argus*, 18 March, viewed 3 April 2017. http://trove.nla.gov.au/ndp/del/article/23233958.

9: The Homogenisation of Rutherglen

Braganza, K 2011, 'The greenhouse effect is real: here's why', *The Conversation*, viewed 28 July 2016, https://theconversation.com/the-greenhouse-effect-is-real-heres-why-1515.

Bureau of Meteorology 2012, 'Australian Climate Observations Reference Network – Surface Air Temperature (ACORN-SAT): Station catalogue', viewed 14 May 2017, http://www.bom.gov.au/climate/change/acorn-sat/documents/ACORN-SAT-Station-Catalogue-2012-WEB.pdf.

Bureau of Meteorology 2014, 'ACORN-SAT Station Adjustment Summary – Rutherglen', viewed 1 July 2016, http://www.bom.gov.au/climate/change/acorn-sat/documents/station-adjustment-summary-Rutherglen.pdf.

Hunt, G 2015, 'Interview: Greg Hunt, Environment Minister', *Lateline*, viewed 2 March 2017, http://www.abc.net.au/lateline/content/2015/s4319361.htm.

Kellow, A 2007, *Science and Public Policy: The Virtuous Corruption of Virtual Environmental Science*, Edward Elgar, Cheltenham.

Lockart, N, Kavetski, D & Franks, SW 2009, 'On the recent warming in the Murray-Darling Basin: Land surface interactions misunderstood', *Geophysical Research Letters*, vol. 36, iss. 24.

Marohasy, J 2016, 'Temperature change at Rutherglen in South-East Australia', *New Climate*, http://dx.doi.org/10.22221/nc.2016.001.

Marohasy, J & Abbot, J 2015, 'Assessing the quality of eight different maximum temperature time series as input when using artificial neural networks to forecast monthly rainfall at Cape Otway, Australia', *Atmospheric Research*, vol. 166, pp. 141–149.

Marohasy, J, Abbot, J, Stewart, K & Jensen, D 2014, 'Modelling Australian and Global Temperatures: What's Wrong? Bourke and Amberley as Case Studies', *The Sydney Papers Online*, iss. 26.

Ryan, TP 1989, *Statistical methods for Quality Improvement*, 3rd edn, John Wiley & Sons, New York.

Taylor, WA 1991, *Optimization & Variation Reduction in Quality*, McGraw-Hill, New York.

10: Moving in Unison: Maximum Temperatures from Victoria, Australia

Brohan, P, Kennedy, JJ, Harris, I, Tett, SFB & Jones, PD 2006, 'Uncertainty estimates in regional and global observed temperature changes: A new data set from 1850', *Journal of Geophysical Research*, vol. 111, iss. D12.

REFERENCES

Bureau of Meteorology 2015, Australian Climate Observations Reference Network – Surface Air Temperature (ACORN-SAT)', Response to the Recommendations of the Technical Advisory Forum, Commonwealth of Australia, Melbourne.

Bureau of Meteorology 2016, 'Annual climate statement 2016', viewed 2 March 2017, http://www.bom.gov.au/climate/current/annual/aus/.

Coates, L, Haynes K, O'Brien, J, McAneney, J & de Oliveira, FD 2014, 'Exploring 167 years of vulnerability: An examination of extreme heat events in Australia 1844–2010', *Environmental Science and Policy*, vol. 42, pp. 33-44.

Deacon, EL 1953, 'Climate Change in Australia since 1880', *Australian Journal of Physics*, vol. 6, pp. 209-218.

Della-Marta, PM, Collins, D & Braganza, K 2004, 'Updating Australia's high-quality annual temperature dataset', *Australian Meteorological Magazine*, vol. 53, iss. 2, pp. 75-93.

Lewis, SC & Karoly, DJ 2014, 'Explaining Extreme Events of 2013: From a Climate Perspective', in a Special Supplement to the *Bulletin of the American Meteorological Society*, vol. 95, no. 9.

Marohasy, J 2016, 'Temperature change at Rutherglen in South-East Australia', *New Climate*, viewed 2 March 2017, http://dx.doi.org/10.22221/nc.2016.001.

Marohasy, J & Abbot, JW 2015, 'Assessing the quality of eight different maximum temperature time series as input when using artificial neural networks to forecast monthly rainfall at Cape Otway, Australia', *Atmospheric Research*, vol. 166, pp. 141-149.

Marohasy, J & Abbot, JW 2016, 'Southeast Australian Maximum Temperature Trends, 1887–2013, in An Evidence-Based Reappraisal', *Evidence-Based Climate Science*, 2nd edn, D Easterbrook (ed), pp. 83-99.

Torok, SJ & Nicholls, N 1996, 'A historical annual temperature dataset for Australia', *Australian Meteorological Magazine*, vol. 45, pp. 251-260.

Trewin, B 2012, 'Techniques involved in developing the Australian Climate Observations Reference Network – Surface Air Temperature (ACORN-SAT) dataset, CAWCR Technical Report No. 049', Bureau of Meteorology, Commonwealth of Australia, Melbourne.

Trewin, B 2013, 'A daily homogenized temperature data set for Australia', *International Journal of Climatology*, vol. 33, iss. 6, pp. 1510-1529.

Vlok, JD 2017, 'Analysis of historical temperature data of Victoria', University of Tasmania Technical report, http://eprints.utas.edu.au/23461/.

11: A Brief Review of the Sun-Climate Connection, with a New Insight Concerning Water Vapour

Ammann, CM, Joos, F, Schimel, DS, Otto-Bliesner, BL & Tomas RA 2007, 'Solar influence on climate during the past millennium: Results from transient simulations with the NCAR Climate System Model', *PNAS*, vol. 104, pp. 3713-3718.

Arguez, A & Vose, RS 2011, 'The definition of the standard WMO climate normal', *Bulletin of the American Meteorological Society*, vol. 92, pp. 699-704.

Asmerom, Y, Polyak, VJ, Rasmussen, JBT, Burns, SJ & Lachniet, M 2013, 'Multidecadal to multicentury scale collapses of Northern Hemisphere monsoons over the past millennium', *PNAS*, vol. 111, pp. 9651-9656.

Bard, E, Raisbeck, G, Yiou, F & Jouzel, J 2000, 'Solar irradiance during the last 1200 years based on cosmogenic nuclides', *Tellus*, vol. 52, iss. 3, pp. 985-992.

Bludova, NG, Obridko, VN & Badalyan, OG 2014, 'The relative umbral area in spot groups as an index of cyclic variation of solar activity', *Solar Physics*, vol. 289, pp. 1013-1028.

Connolly, R & Connolly, M 2014a, 'Urbanization bias I. Is it a negligible problem for global temperature estimates?', *Open Peer Review Journal*, vol. 28 (*Climate Science*) ver. 0.2, viewed 3 February 2015, http://oprj.net/articles/climate-science/28.

Connolly R & Connolly, M 2014b, 'Urbanization bias II. An assessment of the NASA GISS urbanization adjustment method', *Open Peer Review Journal*, vol. 31 (*Climate Science*) ver. 0.1, viewed 27 December 2014, http://oprj.net/articles/climate-science/31.

Connolly R & Connolly, M 2014c, 'Urbanization bias III. Estimating the extent of bias in the Historical Climatology Network datasets', *Open Peer Review Journal*, vol. 34 (*Climate Science*) ver. 0.1, viewed 27 December 2014, http://oprj.net/articles/climate-science/34.

Courtillot, V, Le Mouël, J-L, Blanter, E & Shnirman, M 2010, 'Evolution of seasonal temperature disturbances and solar forcing in the US North Pacific', *Journal of Atmospheric and Solar-Terrestrial Physics*, vol. 72, pp. 83-89.

Cruz-Rico, J, Rivas, D & Tejeda-Martinez, A 2015, 'Variability of surface air temperature in Tampico, northeastern Mexico', *International Journal of Climatology*, vol. 35, pp. 3220-3228.

Essex, C 2011, 'Climate theory versus a theory for climate', *International Journal of Bifurcation and Chaos*, vol. 21, pp. 3477-3487.

Essex, C 2013, 'Does laboratory-scale physics obstruct the development of a theory for climate?', *Journal of Geophysical Research: Atmospheres*, vol. 118, iss. 3, pp. 1218-1225.

Fontenla, JM, Harder, J, Livingston, W, Snow, M & Woods, T 2011, 'High-resolution solar spectral irradiance from extreme ultraviolet to far infrared', *Journal of Geophysical Research: Atmospheres*, vol. 116, iss. D20.

Galindo, S & Saladino, A 2008, 'An early comment on the sunspot-climate connection', *Revista mexicana de física E*, vol. 54, pp. 234-239, viewed May 2015, http://www.scielo.org.mx/pdf/rmfe/v54n2/v54n2a18.pdf.

Gray, LJ, Beer, J, Geller M, Haigh JD, Lockwood, M, Matthes, K, Cubasch, U, Fleitmann, D, Harrison, G, Hood, L, Luterbacher, J, Meehl, GA, Shindell, D & van Geel, B 2010, 'Solar influences on climate', *Review of Geophysics*, vol. 48, iss. 4.

Guttman, NB 1989, 'Statistical descriptors of climate', *Bulletin of the American Meteorological Society*, vol. 70, iss. 6, pp. 602-607.

Harrison, RG & Stephenson, DB 2006, 'Empirical evidence for a nonlinear effect of galactic cosmic rays on clouds', *Proceedings of The Royal Society A*, vol. 462, pp. 1221-1233.

Harrison, RG & Usoskin, I 2010, 'Solar modulation in surface atmospheric electricity', *Journal of Atmospheric and Solar-Terrestrial Physics*, vol. 72, pp. 176-182.

Hathaway, DH 2013, 'A curious history of sunspot penumbrae', *Solar Physics*, vol. 286, pp. 347-356.

Hersbach, H, Peubey, C, Simmons, A, Berrisford, P, Poli, P & Dee, D 2015, 'ERA-20CM: a twentieth-century atmospheric model ensemble', *Quarterly Journal of The Royal Meteorological Society*, vol. 141, pp. 2350-2375.

Herschel, W 1801a, 'Observations tending to investigate the nature of the Sun, in order to find the causes or symptoms of its variable emission of light and heat; with remarks on the use that may possibly be drawn from solar observations', *Philosophical Transactions*, The Royal Society, vol. 91, pp. 265-318.

Herschel, W 1801b, 'Additional observations tending to investigate the symptoms of the variable emission of the light and heat of the Sun; with trials to set aside darkening glasses, by transmitting the solar rays through liquids; and a few remarks to remove objections that might be made against some of the arguments contained in the former paper', *Philosophical Transactions*, The Royal Society, vol. 91, pp. 354-363.

Hoyt, DV 1979, 'Variations in sunspot structure and climate', *Climate Change*, vol. 2, pp. 79-92.

Hoyt, DV & Schatten, KH 1993, 'A discussion of plausible solar irradiance variations, 1700–1992', *Journal of Geophysical Research*, vol. 98, pp. 18895-18906.

Hormes, A, Beer, J & Schluchter, C 2006, 'A geochronological approach to understanding the role of solar activity on Holocene glacier length variability in the Swiss Alps', *Geografiska Annaler*, vol. 88, iss. 4, pp. 281-294.

Kirkby, J, Curtius, J, Almeida, E et al. 2011, 'Role of sulphuric acid, ammonia and galactic cosmic rays in atmospheric aerosol nucleation', *Nature*, vol. 476, pp. 429-433.

Kirkby, J, Duplissy, J, Sengupta, K et al. 2016, 'Ion-induced nucleation of pure biogenic particles', *Nature*, vol. 533, pp. 521-526.

Kobayashi, S, Ota, Y, Harada, Y et al. 2015, 'The JRA-55 reanalysis: General specifications and basic characteristics', *Journal of the Meteorological Society of Japan*, vol. 93, pp. 5-48.

Kossobokov, V, Courtillot, V & Le Mouël, J-L 2010, 'A statistically significant signature of multi-decadal solar activity changes in atmospheric temperatures at three European stations', *Journal of Atmospheric and Solar-Terrestrial Physics*, vol. 72, pp. 595-606.

Krivova, NA, Balmaceda, L & Solanki, SK 2007, 'Reconstruction of solar total irradiance since 1700 from the surface magnetic flux', *Astronomy & Astrophysics*, vol. 467, pp. 335-346.

Krivova, NA, Vieira, LEA & Solanki, SK 2010, 'Reconstruction of solar spectral irradiance since the Maunder minimum', *Journal of Geophysical Research*, vol. 115.

Lam, MM & Tinsley, BA 2016, Solar wind-atmospheric electricity-cloud microphysics connections to weather and climate', *Journal of Atmospheric and Solar-Terrestrial Physics*, vol. 149, pp. 277-290.

Landsberg, HE 1972, 'Weather "normals" and normal weather', NOAA Environmental Data Service, October, pp. 8-13.

Le Mouël, J-L, Blanter, E, Shnirman, M & Courtillot, V 2009, 'Evidence for solar forcing in variability of temperatures and pressures in Europe', *Journal of Atmospheric and Solar-Terrestrial Physics*, vol. 71, pp. 1309-1321.

Lean, J, Beer, J & Bradley, R 1995, 'Reconstruction of solar irradiance since 1610: Implications for climate change', *Geophysical Research Letters*, vol. 22, pp. 3195-3198.

Li, JP & Wang, SH 2008, 'Some mathematical and numerical issues in geophysical fluid dynamics and climate dynamics', *Communications in Computational Physics*, vol. 3, pp. 759-793.

Lions, J-L, Temam, R & Wang, SH 1993, 'Models and numerical analysis of coupled atmosphere-ocean models', *Computational Mechanics Advances*, vol. 1, pp. 1-12.

Livingston, W & Watson, F 2015, 'A new solar signal: Average maximum sunspot magnetic fields independent of activity cycle', *Geophysical Research Letters*, vol. 42, pp. 9185-9189.

Maunder, ASD & Maunder, EW 1908, *The Heavens and Their Story*, Dana Estes and Company, Boston MA.

Menzel, DH 1959, *Our Sun*, Harvard University Press, Cambridge MA.

Miyahara, H, Yokoyama, Y & Masuda, K 2008, 'Possible link between multi-decadal climate cycles and periodic reversals of solar magnetic field polarity', *Earth and Planetary Science Letters*, vol. 272, pp. 290-295.

Monin, AS & Shishkov, YuA 2000, 'Climate as a problem of physics', *Physics-Uspekhi*, vol. 43, iss. 4, pp. 381-406.

Nicoll, KA & Harrison, RG 2014, 'Detection of lower tropospheric responses to solar energetic particles at midlatitudes', *Physical Review Letters*, vol. 112, p. 22501.

Palmer, TN, Doring, A & Seregin, G 2014, 'The real butterfly effect', *Nonlinearity*, vol. 27, no. 9.

Poli, P, Hersbach, H, Dee, D et al. 2016, 'ERA-20C: an atmospheric reanalysis of the twentieth century', *Journal of Climate*, vol. 29, pp. 4083-4097.

Prikryl, P, Iwao, K, Muldrew, DB, Rusin, V, Rybansky, M & Bruntz, R 2016, 'A link between high-speed solar wind streams and explosive extratropical cyclones', *Journal of Atmospheric and Solar-Terrestrial Physics*, vol. 149, pp. 219-231.

Scafetta, N & Willson, RC 2014, 'ACRIM total solar irradiance satellite composite validation versus TSI proxy models', *Astrophysics and Space Science*, vol. 350, pp. 421-442.

Schmidt, GA, Jungclaus, JH, Ammann, CM et al. 2011, 'Climate forcing reconstructions for use in PMIP simulations of the last millennium (v1.0)', *Geoscientific Model Development*, vol. 4, pp. 33-45.

Schmidt, GA, Jungclaus, JH, Ammann, CM et al. 2012, 'Climate forcing reconstructions for use in PMIP simulations of the last millennium (v1.1)', *Geoscientific Model Development*, vol. 5, pp. 185-191.

Schnerr, RS & Spruit, HC 2011, 'The brightness of magnetic field concentrations in the quiet Sun', *Astronomy and Astrophysics,* vol. 532.

Schuhle, U, Wilhelm, K, Hollandt, J, Lemaire, P & Pauluhn, A 2000, 'Radiance variations of the quiet Sun at far-ultraviolet wavelengths', *Astronomy and Astrophysics,* vol. 354, pp. 71-74.

Shapiro, AI, Schmutz, W, Rozanov, E et al. 2011, 'A new approach to the long-term reconstruction of the solar irradiance leads to a large historical solar forcing', *Astronomy and Astrophysics,* vol. 529.

Shaviv, NJ 2005, 'On climate response to changes in the cosmic ray flux and radiative budget', *Journal of Geophysical Research,* vol. 110.

Solheim, J-E, Stordahl, K & Humlum, O 2012, 'The long sunspot cycle 23 predicts a significant temperature decrease in cycle 24', *Journal of Atmospheric and Solar-Terrestrial Physics,* vol. 80, pp. 267-284.

Soon, WWH 2005, 'Variable solar irradiance as a plausible agent for multidecadal variations in the Arctic-wide surface air temperature record of the past 130 years', *Geophysical Research Letters,* vol. 32.

Soon, WWH 2009, 'Solar Arctic-mediated climate variation on multidecadal to centennial timescales: Empirical evidence, mechanistic explanation and testable consequences', *Physical Geography,* vol. 30, pp. 144-184.

Soon, WWH 2014, 'Sun shunned. In Climate Change – The Facts 2014', Institute of Public Affairs, Melbourne, Australia, pp. 57-66.

Soon, W, Baliunas, S, Posmentier, ES, et al. 2000, 'Variations of solar coronal hole area and terrestrial lower tropospheric air temperature from 1979 to mid-1998: astronomical forcings of change in Earth's climate?', *New Astronomy,* vol. 4, pp. 563-579.

Soon, W, Connolly, R & Connolly, M 2015, 'Re-evaluating the role of solar variability on Northern Hemisphere temperature trends since the 19th century', *Earth-Science Reviews,* vol. 150, pp. 409-452.

Soon, W, Dutta, K, Legates, DR et al. 2011, 'Variation in surface air temperature of China during the 20th century', *Journal of Atmospheric and Solar-Terrestrial Physics,* vol. 73, pp. 2331-2344.

Soon, W & Legates, DR 2013, 'Solar irradiance modulation of equator-to-pole (Arctic) temperature gradients: Empirical evidence for climate variation on multi-decadal timescales', *Journal of Atmospheric and Solar-Terrestrial Physics,* vol. 93, pp. 45-56.

Soon, W, Velasco Herrera, VM, Selvaraj, K et al. 2014, 'A review of Holocene solar-linked climatic variation on centennial to millennial timescales: Physical processes, interpretative frameworks and a new multiple cross-wavelet transform algorithm', *Earth-Science Reviews*, vol. 134, pp. 1-15.

Soon, WWH & Yaskell, SH 2003, *The Maunder Minimum and the Variable Sun-Earth Connection*, World Scientific Publishing Company, Singapore.

Steinhilber, F, Beer, J & Fröhlich, C 2009, 'Total solar irradiance during the Holocene', *Geophysical Research Letters*, vol. 36.

Stenflo, JO 2012, 'Scaling laws for magnetic fields on the quiet Sun', *Astronomy and Astrophysics*, vol. 541.

Stenflo, JO & Kosovichev, AG 2012, 'Bipolar magnetic regions on the Sun: Global analysis of the *SOHO*/MDI data set', *The Astrophysical Journal*, vol. 745, p. 129.

Svensmark, H, Enghoff, MB & Pedersen, JOP 2013, 'Response of cloud condensation nuclei (> 50 nm) to changes in ion-nucleation', *Physics Letters A*, vol. 377, pp. 2343-2347.

Svensmark, H & Friis-Christensen, E 1997, 'Variation of cosmic ray flux and global cloud coverage-a missing link in solar-climate relationships', *Journal of Atmospheric and Solar-Terrestrial Physics*, vol. 59, pp. 1225-1232.

Tinsley, B 2012, 'A working hypothesis for connections between electrically-induced changes in cloud microphysics and storm vorticity, with possible effects on circulation', *Advances in Space Research*, vol. 50, pp. 791-805.

Trujillo Bueno, J, Shchukina, N & Asensio Ramos, A (2004) A substantial amount of hidden magnetic energy in the quiet Sun', *Nature*, vol. 430, pp. 326-329.

Usoskin, IG, Voiculescu, M, Kovaltsov, GA & Mursula, K 2006, 'Correlation between clouds at different altitudes and solar activity: Facts or artifact?' *Journal of Atmospheric and Solar-Terrestrial Physics*, vol. 68, pp. 2164-2172.

van Geel, B & Ziegler, PA 2013, 'IPCC underestimates the Sun's role in climate change', *Energy and Environment*, vol. 24, pp. 431-453.

van Loon, H, Brown, J & Milliff, RF 2012, 'Trends in sunspots and North Atlantic sea-level pressure', *Journal of Geophysical Research*, vol. 117.

Veretenenko, S & Ogurtsov, M 2014, 'Stratospheric polar vortex as a possible reason for temporal variations of solar activity and galactic cosmic ray effects on the lower atmospheric circulation', *Advances in Space Research*, vol. 54, pp. 2467-2477.

Vieira, LEA, Solanki, SK, Krivova, NA & Usoskin, I 2011, 'Evolution of the solar irradiance during the Holocene', *Astronomy & Astrophysics*, vol. 531.

Voiculescu, M & Usoskin, I 2012, 'Persistent solar signatures in cloud cover: Spatial and temporal analysis', *Environmental Research Letters*, vol. 7.

Wang, Y-M, Lean, JL & Sheeley, NR Jr 2005, 'Modeling the Sun's magnetic field and irradiance since 1713', *The Astrophysical Journal*, vol. 625, pp. 522-538.

Yan, H, Wei, W, Soon, W, An, ZS, Zhou, W, Liu, Z, Wang, Y & Carter, RM 2015, 'Dynamics of the intertropical convergence zone over the western Pacific during the Little Ice Age', *Nature Geoscience*, vol. 8, pp. 315-320.

Yu, F 2002, 'Altitude variations of cosmic ray induced production of aerosols: Implications for global cloudiness and climate', *Journal of Geophysical Research*, vol. 107.

Yu, F & Luo, G 2014, 'Effect of solar variations on particle formation and cloud condensation nuclei', *Environmental Research Letters*, vol. 9.

Zhao, XH & Feng, XS 2014, 'Periodicities of solar activity and the surface-temperature variation of the Earth and their correlations (in Chinese)', *Chinese Science Bulletin*, vol. 59, pp. 1284-1292.

Zhou, L, Tinsley, BA & Huang, J 2014, 'Effects on winter circulation of short and long term solar wind changes', *Advance Space Research*, vol. 54, pp. 2478-2490.

12: The Advantages of Satellite-Based Regional and Global Temperature Monitoring

Christy, JR, Spencer, RW & Braswell, WD 2000, 'MSU tropospheric temperatures: Dataset construction and radiosonde comparisons', *Journal of Atmospheric and Oceanic Technology*, vol. 17, pp. 1153–1170.

Earth System Science Center, The University of Alabama 2017, 'Global Temperature Report: February', viewed 6 April, 2017, http://www.nsstc.uah.edu/climate/.

Mears, CA & Wentz, FJ 2009, 'Construction of the Remote Sensing Systems V3.2 Atmospheric Temperature Records from the MSU and AMSU Microwave Sounders', *Journal of Atmospheric and Oceanic Technology*, vol. 26, pp. 1040–1056.

Robel, J & Graumann, G (eds.) 2014, *NOAA KLM User's Guide with NOAA-N, N Prime, and MetOp Supplements*, National Oceanic and Atmospheric Administration National Environmental Satellite, Data, and Information Service National Climatic Data Center, Asheville, NC.

Solomon, S, Qin, D, Manning, M, Chen, Z, Marquis, M, Averyt, KB, Tignor M & Miller, HL (eds.) 2007, *Contribution of Working Group I to the Fourth Assessment Report of the Intergovernmental Panel on Climate Change*, Cambridge University Press, Cambridge, United Kingdom and New York, NY, USA.

Spencer, RW & Braswell, WD 2014, 'The role of ENSO in global ocean temperature changes during 1955–2011 simulated with a 1D climate model', *Asia-Pacific Journal of Atmospheric Sciences*, vol. 50, pp. 229–237.

Spencer, RW & Christy, JR 1990, 'Precise monitoring of global temperature trends from satellites'. *Science*, vol. 247, pp. 1558–1562.

Wentz, FJ & Schabel, M 1998, 'Effects of satellite orbit decay on MSU lower tropospheric temperature trends', *Nature*, vol. 394, pp. 361–364.

13: Carbon Dioxide and Plant Growth

Allen, LH Jr, Boote, KJ, Jones, JW, Jones, PH, Valle, RR, Acock, B, Rogers, HH & Dahlman, RC 1987. 'Response of vegetation to rising carbon dioxide: Photosynthesis, biomass, and seed yield of soybean', *Global Biogeochemical Cycles*, vol. 1, no. 1, pp. 1–14.

Ballantyne, AP, Alden, CB, Miller, JB, Tans, PP & White, JW 2012, 'Increase in observed net carbon dioxide uptake by land and oceans during the past 50 years', *Nature*, vol. 488, pp. 70–72.

Bjorkman, O, Badger, M & Armond, PA 1978, 'Thermal acclimation of photosynthesis: Effect of growth temperature on photosynthetic characteristics and components of the photosynthetic apparatus in *Nerium oleander*', *Carnegie Institution of Washington Yearbook*, vol. 77, pp. 262–276.

Center for the Study of Carbon Dioxide and Global Change n.d., 'Percent Dry Weight (Biomass) Increases for 300, 600 and 900 ppm Increases in the Air's CO_2 Concentration', viewed 6 April, 2017, http://www.co2science.org/data/plant_growth/dry/o/oryzas.php.

Ceulemans, R & Mousseau, M 1994, 'Effects of elevated atmospheric CO_2 on woody plants', *New Phytologist*, vol. 127, pp. 425–446.

De Costa, WAJM, Weerakoon, WMW, Chinthaka, KGR, Herath, HMLK & Abeywardena, RMI 2007, 'Genotypic variation in the response of rice (*Oryza sativa* L.) to increased atmospheric carbon dioxide and its physiological basis', *Journal of Agronomy & Crop Science*, vol. 193, pp. 117–130.

Edwards, CA 1997, *Earthworm ecology*, St. Lucie Press, Boca Raton.

Gifford, RM 1979, 'Growth and yield of CO_2-enriched wheat under water-limited conditions', *Australian Journal of Plant Physiology*, vol. 6, pp. 367–378.

Harley, PC, Tenhunen, JD & Lange, OL 1986, 'Use of an analytical model to study the limitations on net photosynthesis in *Arbutus unedo* under field conditions', *Oecologia*, vol. 70, pp. 393–401.

Idso, CD 2012, *The State of Earth's Terrestrial Biosphere: How is it Responding to Rising Atmospheric CO2 and Warmer Temperatures?* Center for the Study of Carbon Dioxide and Global Change, Tempe, AZ, viewed 22 February 2017, http://www.co2science.org/education/reports/greening/TheStateofEarthsTerrestrialBiosphere.pdf.

Idso, CD 2013, *The Positive Externalities of Carbon Dioxide: Estimating the Monetary Benefits of Rising Atmospheric CO2 Concentrations on Global Food Production*, Center for the Study of Carbon Dioxide and Global Change, Tempe, AZ, viewed 22 February 2017, http://www.co2science.org/education/reports/co2benefits/MonetaryBenefitsofRisingCO2onGlobalFoodProduction.pdf.

Idso, CD & Idso, KE 2000, 'Forecasting world food supplies: The impact of the rising atmospheric CO_2 concentration', *Technology*, vol. 7S, pp. 33–55.

Idso, CD & Idso, SB 2011, *The Many Benefits of Atmospheric CO2 Enrichment*, Vales Lakes Publishing, LLC, Pueblo West, Colorado, USA.

Idso, CD & Idso, SB, Carter, RM & Singer, SF (eds.) 2014, *Climate Change Reconsidered II: Biological Impacts,* Chicago, IL: The Heartland Institute.

Idso, KE & Idso, SB 1994, 'Plant responses to atmospheric CO_2 enrichment in the face of environmental constraints: A review of the past 10 years' research', *Agricultural and Forest Meteorology*, vol 69, pp. 153–203.

Idso, SB 1991, 'Carbon dioxide and the fate of Earth', *Global Environmental Change*, vol. 1, pp. 178–182.

Idso, SB 1992, 'Carbon dioxide and global change: End of nature or rebirth of the biosphere?' in: JH Lehr (ed.) *Rational readings on environmental concerns*, Van Nostrand Reinhold, New York, pp. 414–433.

Idso, SB 1995, *CO2 and the Biosphere: The Incredible Legacy of the Industrial Revolution*, Department of Soil, Water & Climate, University of Minnesota, St. Paul, Minnesota, USA.

Loehle, C, Idso, C & Wigley, TB 2016, 'Physiological and ecological factors influencing recent trends in United States forest health responses to climate change', *Forest Ecology and Management*, vol. 363, pp. 179–189.

Jurik, TW, Weber, JA & Gates, DM 1984, 'Short-term effects of CO_2 on gas exchanges of leaves of bigtooth aspen (*Populus grandidentata*) in the field', *Plant Physiology*, vol. 75, pp. 1022–1026.

Loehle, C, Idso, C & Wigley, TB 2016, 'Physiological and ecological factors influencing recent trends in United States forest health responses to climate change', *Forest Ecology and Management*, vol. 363, pp. 179–189.

Long, SP 1991, 'Modification of the response of photosynthetic productivity to rising temperature by atmospheric CO_2 concentrations: Has its importance been underestimated?' *Plant, Cell and Environment*, vol. 14, pp. 729–739.

McMurtrie, RE, Comins, HN, Kirschbaum, MUF & Wang, Y-P 1992, 'Modifying existing forest growth models to take account of effects of elevated CO_2', *Australian Journal of Botany*, vol. 40, pp. 657–677.

Nilsen, S, Hovland, K, Dons, C & Sletten, SP 1983, 'Effect of CO_2 enrichment on photosynthesis, growth and yield of tomato', *Scientia Horticulturae*, vol. 20, pp. 1–14.

Poorter, H 1993, 'Interspecific variation in the growth response of plants to an elevated ambient CO_2 concentration', *Vegetatio*, vol. 104–105, pp. 77–97.

Rogers, HH, Runion, GB & Krupa, SV 1994, 'Plant responses to atmospheric CO_2 enrichment with emphasis on roots and the rhizosphere', *Environmental Pollution*, vol. 83, pp. 155–189.

Seemann, JR, Berry, JA & Downton, JS 1984, 'Photosynthetic response and adaptation to high temperature in desert plants. A comparison of gas exchange and fluorescence methods for studies of thermal tolerance', *Plant Physiology*, vol. 75, pp. 364–368.

Stocker, TF, Quin, D, Plattner, GK, Tignor, M, Allen, SK, Boschung, J, Nauels, A, Xia, Y, Bex, Midgely, PM (eds.) 2013, *Climate Change 2013: The Physical Science Basis, The Intergovernmental Panel on Climate Change (IPCC)*, Cambridge University Press, Cambridge, United Kingdom and New York, NY, USA.

Stuhlfauth, T & Fock, HP 1990, 'Effect of whole season CO_2 enrichment on the cultivation of a medicinal plant, *Digitalis lanata*', *Journal of Agronomy and Crop Science*, vol. 164, pp. 168–173.

Tans, P 2009 'An accounting of the observed increase in oceanic and atmospheric CO_2 and an outlook for the future', *Oceanography*, vol. 22, pp. 26–35.

Trimble, SW & Crosson, P 2000, 'U.S. soil erosion rates – myth and reality', *Science*, vol. 289, pp. 248–250.

Wallace, A 1994, 'Soil organic matter is essential to solving soil and environmental problems', *Communications in Soil Science and Plant Analysis*, vol. 25, pp. 15–28.

Wittwer, SH 1982, 'Carbon dioxide and crop productivity', *New Scientist*, vol. 95, pp. 233–234.

Wittwer, SH 1995, *Food, climate, and carbon dioxide: The global environment and world food production*, Lewis publishers, Boca Raton.

Wittwer, SH 1997, 'The global environment: It's good for food production' in: PJ Michaels (ed.), *State of the climate report: Essays on global climate change,* New Hope Environmental Services, New Hope, pp. 8–13.

Wullschleger, SD, Post, WM & King, AW 1995, 'On the potential for a CO_2 fertilization effect in forests: Estimates of the biotic growth factor based on 58 controlled-exposure studies', in: GM Woodwell & FT Mackenzie (eds.), *Biotic Feedbacks in the Global Climate System,* Oxford University Press, Oxford, pp. 85–107. 14: The Poor Are Carrying the Cost of Today's Climate Policy

14: The Poor Are Carrying the Cost of Today's Climate Policy

Ahlenius, H 2006, 'The greening of Niger', *World Resources 2008: Roots of Resilience – Growing the Wealth of the Poor,* viewed 6 March 2017, http://www.wri.org/publication/world-resources-2008.

Bryce, R 2010, 'Despite billions in subsidies, corn ethanol has not cut U.S. oil imports', *Issue Brief,* no. 7, Manhattan Institute for Policy Research, viewed 3 March 2017, http://www.manhattan-institute.org/pdf/ib_07.pdf.

Dlouhy, JA 2016, 'As Corn Devours U.S. Prairies, Greens Reconsider Biofuel Mandate', *Bloomberg Politics,* 27 July, viewed 3 Mach, 2017, https://www.bloomberg.com/politics/articles/2016-07-27/as-corn-devours-u-s-prairies-greens-reconsider-biofuel-mandate.

D'Olivo, JP, McCulloch, NT & Judd, K 2013, 'Long-term records of coral calcification across the central Great Barrier Reef: assessing the impacts of river runoff and climate change' *Coral Reefs,* vol. 32, pp. 999–1012.

Ernstig, A 2015, 'Drax in the UK: subsidies for burning coal and increasingly more and more wood from overseas', *Biofuelwatch,* viewed 3 March 2017, http://globalforestcoalition.org/wp-content/uploads/2015/01/case-study-7_almuth_UK.pdf.

Field, CB, Barros, VR, Dokken, DJ, Mach, KJ, Mastrandrea, MD,Bilir, TE, Chatterjee, M, Ebi, KL, Estrada, YO, Genova, RC, Girma, B, Kissel, ES, Levy, AN, MacCracken, S, Mastrandrea, PR & White, LL (eds.) 2014, '*Climate Change 2014: Impacts, Adaptation, and Vulnerability', The Intergovernmental Panel on Climate Change (IPCC),* Cambridge University Press, Cambridge, United Kingdom and New York, NY, USA.

Fisher, T & Fitzsimmons, A 2013, 'Big Wind's Dirty Little Secret: Toxic Lakes and Radioactive Waste', Institute for Energy Research, viewed 3 March 2017, http://instituteforenergyresearch.org/analysis/big-winds-dirty-little-secret-rare-earth-minerals/.

Food and Agriculture Organization 2009, 'What happened to world food prices and why?' viewed 3 March 2017, ftp://ftp.fao.org/docrep/fao/012/i0854e/i0854e01.pdf.

Fukuda, SY, Suzuki, Y & Shiraiwa, Y 2014, 'Difference in physiological responses of growth, photosynthesis and calcification of the coccolithophore *Emiliania huxleyi* to acidification by acid and CO2 enrichment', *Photosynthesis Research*, vol. 121, pp. 299–309.

Gado, S 2015, 'The Energy Sector of Nigeria: Perspectives and Opportunities', energy Charter Secretariat, Brussels, Belgium, viewed 6 March 2017, http://www.energycharter.org/fileadmin/DocumentsMedia/Occasional/Niger_Energy_Sector.pdf.

Gosden, E 2016, 'EU drive for "green" biodiesel has increased emissions, study finds', *The Daily Telegraph*, 26 April, viewed 3 March 2017, http://www.thegwpf.com/eu-drive-for-green-biodiesel-has-increased-emissions-study-finds/#sthash.b4yTTbzg.dpuf.

Hao, Z, AghaKouchak, A, Nakhjiri, N & Farahmand A 2014, 'Global integrated drought monitoring and prediction system', *Nature*, viewed 22 February 2017, http://www.nature.com/articles/sdata20141.

Hughes, G 2012, 'The impact of wind power on household energy bills: Evidence to the House of Commons Energy and Climate Change Committee', The Global Warming Policy Foundation, viewed 3 March 2017, http://www.thegwpf.org/images/stories/gwpf-reports/hughes-evidence.pdf.

Idso, CD 2013, *The Positive Externalities of Carbon Dioxide: Estimating the Monetary Benefits of Rising Atmospheric CO2 Concentrations on Global Food Production*, Center for the Study of Carbon Dioxide and Global Change, Tempe, AZ, viewed 22 February 2017, http://www.co2science.org/education/reports/co2benefits/MonetaryBenefitsofRisingCO2onGlobalFoodProduction.pdf.

Idso, CD & Idso, SB 2011, *The Many Benefits of Atmospheric CO2 Enrichment*, Vales Lakes Publishing, LLC, Pueblo West, Colorado, USA.

International Renewable Energy Agency 2013, 'Niger: Renewables Readiness Assessment 2013', viewed 6 March 2017, http://www.irena.org/DocumentDownloads/Publications/RRA_Niger.pdf.

McCarthy, A, Rogers, SP, Duffy, SJ & Campbell, DA 2012, 'Elevated carbon dioxide differentially alters the photophysiology of *Thalassiosira pseudonana* (Bacillariophyceae) and *Emiliania huxleyi* (Haptophyta), *Journal of Phycology*, vol. 48, pp. 635–646.

Montford, A 2015, 'Unintended consequences of climate change policy', The Global Warming Policy Foundation, Report 16, viewed 3 March 2017, http://www.thegwpf.org/content/uploads/2015/01/Unintended-Consequences.pdf.

Moss, T & Leo, B 2014, 'Maximizing Access to Energy: Estimates of Access and Generation for the Overseas Private Investment Corporation's Portfolio,' Center for Global Development, viewed 3 March 2017, https://www.cgdev.org/sites/default/files/maximizing-access-energy-opic_1.pdf.

Oerlemans, J 1994, 'Quantifying global warming from the retreat of glaciers', Science, vol. 264, pp. 243–244.

Palacios, S & Zimmerman, RC 2007, Response of eelgrass Zostera marina to CO_2 enrichment: possible impacts of climate change and potential for remediation of coastal habitats, Marine Ecology Progress Series, vol. 344, p. 113.

Public Health England 2014, 'Local action on health inequalities: Fuel poverty and cold home-related health problems', Health Equity Evidence Review 7, viewed 3 March 2014, https://www.gov.uk/government/uploads/system/uploads/attachment_data/file/357409/Review7_Fuel_poverty_health_inequalities.pdf.

Ridley, M 2012, 'The windfarm delusion', The Spectator, 3 March, viewed 3 March 2017, http://www.spectator.co.uk/2012/03/the-winds-of-change/.

Rose, D 2015, 'The UK's £1 billion carbon-belcher raping US forests … that YOU pay for: How world's biggest green power plant is actually INCREASING greenhouse gas emissions and Britain's energy bill', The Mail on Sunday, 7 June, viewed 3 March 2017, http://www.dailymail.co.uk/news/article-3113908/How-world-s-biggest-green-power-plant-actually-INCREASING-greenhouse-gas-emissions-Britain-s-energy-bill.htm.

Smith, HEK, Tyrrell, T, Charalampopoulou, A, Dumousseaud, C, Legge, OJ, Birchenough, S, Pettit, LR, Garley, R, Hartman, SE, Hartman, MC, Sagoo, N, Daniels, CJ, Achterberg, EP & Hydes, DJ 2012, 'Predominance of heavily calcified coccolithophores at low CaCO3 saturation during winter in the Bay of Biscay', PNAS, vol. 109, pp. 8845–8849.

Stephenson, AL & MacKay, DJC 2014, 'Life Cycle Impacts of Biomass Electricity in 2010', Department of Energy & Climate Change, July, p. 35, viewed 3 March 2017, https://www.gov.uk/government/uploads/system/uploads/attachment_data/file/349024/BEAC_Report_290814.pdf.

Webb, AP & Kench, PS 2010, 'The dynamic response of reef islands to sea-level rise: Evidence from multi-decadal analysis of island change in the Central Pacific', Global and Planetary Change, vol. 72, iss. 3, pp. 234–246.

'UCN Red List says global polar bear population is 22,000 – 31,000 (26,000)' 2015, *Polar Bear Science: Past and Present,* viewed 6 March 2017, https://polarbearscience.com/2015/11/18/iucn-red-list-says-global-polar-bear-population-is-22000-31000-26000/.

Yim, S & Barrett, S 2012, 'Public Health Impacts of Combustion Emissions in the United Kingdom', *Environmental Science & Technology,* vol. 46, no. 8, pp. 4291–4296.

Zaichun, Z, Piao, S, Myneni, RB, Huang, M, Zeng, Z, Canadell, JG, Ciais, P, Sitch,S, Friedlingstein, P, Arneth, A, Cao, C, Cheng, L, Kato, E, Koven, C, Li, Y, Lian, X, Liu, Y, Liu, R, Mao, J, Pan, Y, Peng, S, Peñuelas, J, Poulter, B, Pugh, TAM & Stocker, BD 2016, 'Greening of the Earth and its drivers', *Nature Climate Change,* vol. 6, pp. 791–795.

15: The Impact and Cost of the 2015 Paris Climate Summit, with a Focus on US Policies

Böhringer, C, Rutherford, TF & Tol, RSJ (eds.) 2009, 'The EU 20/20/2020 targets: An overview of the EMF22 assessment', *Energy economics,* vol. 31, supplement 2, pp. 268-273.

Boyd, R, Turner, J & Ward, B 2015, 'Tracking intended nationally determined contributions: what are the implications for greenhouse gas emissions in 2030?', Grantham Research Institute on Climate Change and the Environment, viewed 3 March 2017, http://www.lse.ac.uk/GranthamInstitute/publication/tracking-intended-nationally-determined-contributions-what-are-the-implications-for-greenhouse-gas-emissions-in-2030/.

Buckley, C 2015, 'China Burns Much More Coal than Reported, complicating Climate Talks', *The New York Times,* 3 November, viewed 14 September 2016, http://www.nytimes.com/2015/11/04/world/asia/china-burns-much-more-coal-than-reported-complicating-climate-talks.html?_r=0.

Calvin, K, Clarke, L, Krey, V, Blanford, G, Jiang, K, Kainuma, M, Kriegler, E, Luderer, G & Shukla, PR 2012, 'The Role of Asia in Mitigating Climate Change: Results from the Asia Modeling Exercise', *Energy Economics,* vol. 34, The Asia Modeling Exercise: Exploring the Role of Asia in Mitigating Climate Change, supplement 3, pp. S251-60.

Calvin, K, Fawcett, A & Kejun, J 2012, 'Comparing Model Results to National Climate Policy Goals: Results from the Asia Modeling Exercise,' *Energy Economics,* vol. 34, pp. 306-315.

Capros, P, Mantzos, L, Parousos, L, Tasios, N, Klaassen, G & Van Ierland, T 2011, 'Analysis of the EU policy package on climate change and renewables', *Energy Policy*, vol. 39, no. 3, pp. 1476-1485.

China INDC 2015, 'China INDC Submission', viewed 6 March 2017, http://www4.unfccc.int/Submissions/INDC/Published%20Documents/China/1/China's%20INDC%20-%20on%2030%20June%202015.pdf.

Climate Action Tracker 2016a, 'China', viewed 14 September 2016, http://climateactiontracker.org/countries/china.html.

Climate Action Tracker 2016b, 'Mexico', viewed 14 September 2016, http://climateactiontracker.org/countries/mexico.html.

Climate Interactive. 2015. 'Data for INDC and Baseline.' https://d168d9ca7ixfvo.cloudfront.net/wp-content/uploads/2013/12/Climate-Scoreboard-Output-27Oct2015-to-share.xlsx.

Clinton, WJ 1993, 'Remarks on Earth Day.' 21 April, viewed 16 September 2016, http://www.presidency.ucsb.edu/ws/?pid=46460.

Energy Information Administration (EIA) 2016, 'Annual Energy Outlook 2016 with projections to 2040', Data for Annual Energy Outlook 2015, viewed 16 August 2016, http://www.eia.gov/outlooks/aeo/pdf/0383(2016).pdf.

EU INDC 2015, 'European Union INDC Submission,' viewed 14 September 2016, http://www4.unfccc.int/submissions/INDC/Published%20Documents/Latvia/1/LV-03-06-EU%20INDC.pdf.

Fawcett, AA, Clarke, LC, Rausch, S & Weyant, JP 2013, 'Overview of EMF 24 policy scenarios', *The Energy Journal*, vol. 25, no. 1, pp. 33-60.

IPCC 2007, *Simple Climate Models. Climate Change 2007: Working Panel Group I: The Physical Science Basis*, viewed 14 September 2016, https://www.ipcc.ch/publications_and_data/ar4/wg1/en/ch8s8-8-2.html.

Edenhofer, O, Pichs-Madruga, R, Sokona, Y, Farahani, E, Kadner, S, Seyboth, K, Adler, A, Baum, I, Brunner, S, Eickemeier, P, Kriemann, B, Savolainen, J, Schlömer, S, von Stechow, C & Zwickel, T (eds.) 2014, *Summary for Policymakers. In: Climate Change 2014: Mitigation of Climate Change. Contribution of Working Group III to the Fifth Assessment Report of the Intergovernmental Panel on Climate Change*, Cambridge University Press, Cambridge, United Kingdom and New York, NY, USA.

Joby, W & Baker, P 1997, 'Clinton Details Global Warming Plan', *Washington Post*, 23 October, viewed 16 September 2016, http://www.washingtonpost.com/wp-srv/inatl/longterm/climate/stories/clim102397.htm.

Knopf, B, Chen, YHH, De Cian, E, Förster, H, Kanudia, A, Karkatsouli, I, Keppo, I, Koljonen, T, Schumacher, K & Van Vuuren, DP 2013, 'Beyond 2020—Strategies and costs for transforming the European energy system', *Climate Change Economics*, vol. 4, no. 1, pp. 1-38.

Kriegler, Elmar, John P. Weyant, Geoffrey J. Blanford, Volker Krey, Leon Clarke, Jae Edmonds, Allen Fawcett, et al. 2014. 'The Role of Technology for Achieving Climate Policy Objectives: Overview of the EMF 27 Study on Global Technology and Climate Policy Strategies.' Climatic Change 123 (3-4): 353–67. doi:10.1007/s10584-013-0953-7.

Lomborg, B 2016, 'Impact of current climate proposals', *Global Policy*, vol. 7, no. 1, pp. 109-118.

MIT 2015, '2015 Energy and Climate Outlook', viewed 6 March 2017, http://globalchange.mit.edu/research/publications/other/special/2015Outlook.

Netherlands EPA 2015, 'PBL Climate Pledge INDC Tool', viewed 6 March 2017, http://infographics.pbl.nl/indc/.

Rodney Boyd, Joe Cranston Turner, and Bob Ward. 2015. 'Intended Nationally Determined Contributions: What Are the Implications for Greenhouse Gas Emissions in 2030?' Accessed November 25. http://www.lse.ac.uk/GranthamInstitute/publication/intended-nationally-determined-contributions-what-are-the-implications-for-greenhouse-gas-emissions-in-2030/.

Tol, RSJ 2012, 'A cost–benefit analysis of the EU 20/20/2020 package', *Energy Policy*, vol. 49, pp. 288-295.

United Nations 1992, 'UN General Assembly Intergovernmental Committee for a Framework Convention on Climate Change', 27 May, viewed 16 September 2016, http://unfccc.int/resource/docs/a/18p2a01c01.pdf.

United Nations 2006, 'Compendium of methods and tools to evaluate impacts of, and vulnerability of adaptation to, climate change', United Nations Framework Convention on Climate Change, viewed 14 September 2015, http://unfccc.int/adaptation/nairobi_work_programme/knowledge_resources_and_publications/items/5430.php.

United Nations 2010, 'Cancun Climate Change Conference – November 2010', viewed 16 September 2016, http://unfccc.int/meetings/cancun_nov_2010/meeting/6266.php.

United Nations Framework Convention on Climate Change (UNFCCC) 2015, 'Synthesis Report on the Aggregate Effect of the Intended Nationally Determined Contributions', viewed 6 March 2017, http://unfccc.int/resource/docs/2015/cop21/eng/07.pdf.

United Nations Environment Programme (UNEP) 2015, 'The Emissions Gap Report 2015', viewed 6 March 2017, http://uneplive.unep.org/media/docs/theme/13/EGR_2015_ES_English_Embargoed.pdf.

US INDC 2015, 'United States INDC Submission,' viewed 6 March 2017, http://www4.unfccc.int/submissions/INDC/Published%20Documents/United%20States%20of%20America/1/U.S.%20Cover%20Note%20INDC%20and%20Accompanying%20Information.pdf.

Veysey, J, Octaviano, C, Calvin, K, Herreras Martinez, S, Kitous, A, McFarland, J & van der Zwaan, B 2016, 'Pathways to Mexico's climate change mitigation targets: A multi-model analysis,' *Energy Economics*, vol. 56, pp. 587-599.

16: Re-examining Papal Energy and Climate Ethics

Bezdek, R 2016, 'Unsung Role of Coal in the Miracle of U.S. Growth: Past, Present and Future, Part 1', *Public Utilities Fortnightly*, August, viewed 13 April, http://misi-net.com/publications/PUF-0816.pdf

Boynes, C 2004, *Voices from Africa: Biotechnology and the subsistence farmer*, Congress of Racial Equality, New York.

Bryce, R 2014, 'Not beyond coal: How global thirst for low-cost electricity continues driving coal demand', Manhattan Institute, October 27, viewed 7 March 2017, https://www.manhattan-institute.org/html/not-beyond-coal-how-global-thirst-low-cost-electricity-continues-driving-coal-demand-5991.html.

Bryce, R2015, 'Obama's wind-energy lobby gets blown away: California judge rules in favor of bald eagles and against 30-year permits to shred them', *Wall Street Journal*, August 19.

Carter, RM 2010, *Climate: The Counter Consensus*, London: Stacey International.

Cornwall Alliance 2015, 'Open letter to Pope Francis on climate change', viewed 7 March 2017, http://cornwallalliance.org/anopenlettertopopefrancison climatechange/.

Dasgupta, P, Ramanathan, p. Raven, P, Sorondo, Mgr M, Archer, M, Crutzen, PJ, Lena, P, Lee, YT, Molina, MJ, Rees, M, Sachs, J, Schellnhuber, J & Sorondo, Mgr M (eds.) 2015, 'Climate Change and the Common Good: A statement of the problem and the demand for transformative solutions', Pontifical Academies of Science, Vatican City.

Driessen, P 2014a, *Miracle Molecule: Carbon dioxide, gas of life*, CFACT Books, Washington, DC.

Driessen, P 2014b, "'Climate-Smart' Policies for Africa Are Stupid", Townhall. com, August 9; https://townhall.com/columnists/pauldriessen/2014/08/09/ climatesmart-policies-for-africa-are-stupid-n1876608.

Driessen, P 2005, 'Sustainable development = Sustained poverty: Keeping developing countries cute, indigenous, electricity-poor – and impoverished', *Free Pakistani Newsletter*, August 1.

Durkin, M 1997, *Against Nature*, WAG-TV Production Company and Channel 4 Television.

Feynman, R 1985, 'The Character of Physical Law', transcript of the Messenger Lectures at Cornell University, presented in November 1964, 12th edn, MIT Press, Cambridge, MA.

Friedman, L 2010, 'South Africa wins $3.75 billion coal loan', *New York Times*, April 9.

Heartland Institute 2016, 'Why the Pope is wrong about global warming', viewed 7 March 2017, https://www.heartland.org/topics/climate-change/ tell-pope-francis-global-warming-is-not-a-crisis/.

Heerman, K 2015, 'International economic datasets for 189 countries', US Department of Agriculture, Economic Research Service, viewed 7 March 2017, https://www.ers.usda.gov/data-products/international-macroeconomic-data-set. aspx.

Helman, C 2010, 'Obama doubletalk muddies Ghana energy investment', *Forbes*, February 25.

Iberdrola Renewables 2008, 'Big Horn Wind River Project', Overview and details about 200 MW wind energy project, viewed 13 April 2017, https://s3-us-west-2.amazonaws.com/iberdrola-pdfs/pdf/Big_Horn_fact_sheet.pdf

International Energy Agency, 'Energy Poverty and Energy Access'.

Louw, L 2002, 'Poverty today is truly miraculous', *The Telegraph* (London), January 9.

Moran, A (ed.) 2015, *Climate Change: The Facts*, Melbourne: Institute of Public Affairs.

Morano, M 2015, 'Unholy Alliance: Exposing the radicals advising Pope Francis on climate', Climate Depot, viewed 7 March 2017, http://www.climatedepot. com/2015/09/24/special-report-unholy-alliance-exposing-the-radicals-advising-pope-francis-on-climate/.

NIPCC 2014, *Climate Change Reconsidered II: Biological Impacts*, Heartland Institute, Chicago, IL, pp. 12–160.

Nongovernmental International Panel on Climate Change (NIPCC) 2014, *Climate Change Reconsidered II: Physical Science*, Heartland Institute, Chicago.

Pope Francis 2015, *Laudato Si: on care for our common home*, Encyclical Letter, May 24, viewed 6 March 2017, http://w2.vatican.va/content/francesco/en/encyclicals/documents/papa-francesco_20150524_enciclica-laudato-si.html.

Sagan, C 1980, *Cosmos*, Random House, New York.

Singer FS & Avery DT 2007, *Unstoppable Global Warming: Every 1,500 years*, Rowman & Littlefield Publishing Group, Maryland.

Spencer, RW 2010, *The Great Global Warming Blunder: How Mother Nature fooled the world's top climate scientists*, New York: Encounter Book.

Wiegand, J 2015, ' AWEA's Eagle Mortality Defense: A Response', *Master Resource*, 31 August, viewed 7 March 2017, https://www.masterresource.org/american-wind-energy-association/awea-eagle-rebuttal-bryce/.

World Health Organization (n.d.), 'Coming clean: modern fuels, modern stoves', *Fuel for Life: Household Energy and Health*, pp. 28–29.

Zaimov, S 2015, 'UN and Vatican agree: Man-made climate change is a "scientific reality", call it a "moral and religious imperative for humanity"', *Christian Post*, April 29.

Zillman, C 2015, 'Here's what Pope Francis told Congress to do', *Fortune*, September 24.

17: Free Speech and Climate Change

Adler, JH 2016, 'Making Defamation Law Great Again: Michael Mann's suit may continue', *The Washington Post*, 22 December, viewed 27 March 2017, https://www.washingtonpost.com/news/volokh-conspiracy/wp/2016/12/22/making-defamation-law-great-again-michael-manns-suit-may-continue/?utm_term=.38e120420758.

Alvord, A 2016, 'Deception on climate change products is fraud, not free speech', *Sacramento Bee*, 17 June, viewed 27 March 2017, http://www.sacbee.com/opinion/california-forum/article84202042.html.

Brett, J 2014, 'Must we choose between climate-change action and freedom of speech?' *The Monthly*, August, viewed 27 April 2017, https://www.themonthly.com.au/issue/2014/august/1406815200/judith-brett/must-we-choose-between-climate-change-action-and-freedom.

'CEI Fights Subpoena to Silence Debate on Climate Change' 2016, Competitive Enterprise Institute, 7 April, viewed 27 March 2017, https://cei.org/content/cei-fights-subpoena-silence-debate-climate-change.

Democratic Platform Committee 2016, 2016 Democratic Party Platform, viewed 27 March 2017, http://www.presidency.ucsb.edu/papers_pdf/117717.pdf.

Epstein, RA 2016, 'Does the First Amendment Protect Global Warming Deniers?', *Newsweek*, 22 April, viewed 27 March 2017, http://www.newsweek.com/does-first-amendment-protect-global-warming-deniers-449864.

Flood, A 2016, 'Free-speech group slams Portland schools' ban on books that question climate change', *The Guardian*, 25 May, viewed 27 March 2017, https://www.theguardian.com/books/2016/may/24/portland-schools-ditch-textbooks-that-question-climate-change.

Foran, C 2016, 'Donald Trump and the Triumph of Climate-Change Denial', *The Atlantic*, 25 December, viewed 27 March 2017, https://www.theatlantic.com/politics/archive/2016/12/donald-trump-climate-change-skeptic-denial/510359/.

Hasham, N 2015, 'Students and staff warn of angry backlash if "sceptical environmentalist" Bjorn Lomborg sets up research centre at Flinders University', *The Sydney Morning Herald*, 24 July, viewed 27 March 2017, http://www.smh.com.au/federal-politics/political-news/students-and-staff-warn-of-angry-backlash-if-sceptical-environmentalist-bjorn-lomborg-sets-up-research-centre-at-flinders-university-20150724-gijmlz.html.

Houghton, JT, Jenkins GJ & Ephraums, JJ 1991, *The IPCC Scientific Assessment*, Cambridge University Press, MA.

Kraft, ME 2016, 'Climate-change deniers deserve punishment', *Providence Journal*, 11 April, viewed 27 March 2017, http://www.providencejournal.com/opinion/20160411/michael-e-kraft-climate-change-deniers-deserve-punishment.

Laden, G 2015, 'Letter To President Obama: Investigate Deniers Under RICO', *Greg Laden's Blog* on *Science Blogs*, 19 September, viewed 27 March 2017, http://scienceblogs.com/gregladen/2015/09/19/letter-to-president-obama-investigate-deniers-under-rico/.

Lavik, T 2016, 'Climate change denial, freedom of speech and global justice', *Nordic Journal of Applied Ethics*, vol. 10, no. 2, pp. 75–90.

Markey, EJ, Boxer, B & Whitehouse, S 2015, Letter to Mr Jack Gerard, United States Senate, 25 February, viewed 27 March 2017, https://www.markey.senate.gov/imo/media/doc/02-25-15_Markey_ClimateDenialOversight_AmericanPetroleumInstitute.pdf.

Mooney, K 2016, 'Senate Liberals, Targeting Climate Change 'Deniers,' Demand to Know Donors to 22 Think Tanks', *The Daily Signal*, 3 August, viewed 27 March 2017, http://dailysignal.com/2016/08/03/senate-liberals-targeting-climate-change-deniers-demand-to-know-donors-to-22-think-tanks/.

O'Neill, B 2006, 'Global warming: the chilling effect on free speech', *Spiked*, 6 October, viewed 27 March 2017, http://www.spiked-online.com/newsite/article/1782#.WNiqO1WGPmE.

Ridley, M 2017, 'Climate change: Politics and science are a toxic combination', *The Australian*, 7 February, viewed 27 March 2017, http://www.theaustralian.com.au/news/world/the-times/climate-change-politics-and-science-are-a-toxic-combination/news-story/89da1b787fff9481f2f6ece000b880c4.

Rivkin Jr, DB & Grossman, AM 2016, 'Punishing Climate-Change Skeptics', CATO Institute, March 23, viewed 27 March 2017, https://www.cato.org/publications/commentary/punishing-climate-change-skeptics.

Shearman DJC & Smith JW 2007, *The Climate Change Challenge and the Failure of Democracy*, in C Berg, *Liberty, Equality and Democracy*, Connor Court Publishing, Australia, p. 75.

Sidahmed, M 2016, 'Climate change denial in the Trump cabinet: where do his nominees stand?' *The Guardian*, 16 December, viewed 27 March 2017, https://www.theguardian.com/environment/2016/dec/15/trump-cabinet-climate-change-deniers.

Simberg, R 2012, 'The Other Scandal in Unhappy Valley', Competitive Enterprise Institute, 13 July, viewed 27 March 2017, https://cei.org/blog/other-scandal-unhappy-valley.

Steyn, M 2012, 'Football and Hockey', *National Review Online*, 15 July, viewed 27 March 2017, http://www.nationalreview.com/corner/309442/football-and-hockey-mark-steyn.

Thomson, K 2006, Letter from the Member for Wills, IPA Library, 27 September, viewed 27 March 2107, http://ipa.org.au/library/publication/1273802767_document_thomson_letter.pdf.

Torcello, L 2014, 'Is misinformation about the climate criminally negligent?', The Conversation, 13 March, viewed 27 March 2017, https://theconversation.com/is-misinformation-about-the-climate-criminally-negligent-23111.

Wade, T 2016, 'U.S. states signed pact to keep Exxon climate probe confidential', *Reuters*, 4 August, viewed 27 March 2017, http://www.reuters.com/article/us-exxon-mobil-climatechange-idUSKCN10F0AX.

Woodruff, J 2015, 'Has Exon Mobil misled the public about its climate change research?' PBS Newshour, viewed 27 March 2017, http://www.pbs.org/newshour/bb/exxon-mobil-mislead-public-climate-change-research/.

Worland, J 2017, 'Scientists Plan a March on Washington to Challenge President Trump', *Time*, 25 January, viewed 27 March 2017, http://time.com/4648570/scientists-washington-donald-trump-march-climate-change/.

18: The Lukewarm Paradigm and Funding of Science

Christy, JR 2016, 'Testimony to the U.S. House Committee on Science, Space & Technology', 2 February, viewed 28 March, 2017, http://docs.house.gov/meetings/SY/SY00/20160202/104399/HHRG-114-SY00-Wstate-ChristyJ-20160202.pdf.

Fanelli, D, 2012, 'Negative results are disappearing from most disciplines and countries', *Scienometrics*, vol. 90, pp. 891–904.

Hourdin, F, Mauritsen, T, Gettelman, A, Golaz, J, Balaji, V, Duan, Q, Folini D, Ji, D, Klocke, D, Qian, Y, Rauser, F, Rio, C, Tomassini, L, Watanabe, M, Williamson D 2017, 'The art and science of climate model tuning', Bulletin of the American Meteorological Society, http://journals.ametsoc.org/doi/10.1175/BAMS-D-15-00135.1.

Ioannidis, JP 2005, 'Why most published research findings are false', *PLoS Medicine*, vol 2., pp. e124.

Karl, TR, Arguez, A, Huang, B, Lawrimore, JH, McMahon, JR, Menne, MJ, Peterson, TC, Vose, RS & Zhang, H-M 2015, 'Possible artifacts of data biases in the recent global surface warming hiatus', *Science,* vol. 348, pp. 1469–1472.

Kuhn, T 1996, *The Structure of Scientific Revolutions*, University of Chicago Press.

Lewis, N 2015, 'Implications of lower aerosol forcing for climate sensitivity', *Climate Etc.*, 19 March, viewed 6 April, 2017, https://judithcurry.com/2015/03/19/implications-of-lower-aerosol-forcing-for-climate-sensitivity/.

Michaels PJ & Knappenberger, PC 1996, 'Human effect on global climate?', *Nature*, vol. 384, pp. 522–523.

Michaels, PJ, Knappenberger, PC, Frauenfeld, OW & Davis, RE 2002, 'Revised 21st century temperature projections', *Climate Research*, vol. 23, pp. 1–9.

Michaels, PJ 2015, 'Why climate models are failing', *Climate Change: The Facts*, in A Moran (ed.), Stockade Books, Woodsville.

Santer, BD, Taylor, KE, Wigley, TML, Johns, TC, Jones, PD, Karoly, DJ, Mitchell, JFB, Oort, AH, Penner, JE, Ramaswamy, V, Schwarzkopf, MD, Stouffer, RJ & Tett, S 1996, 'A search for human influences on the thermal structure of the atmosphere', *Nature*, vol. 382, pp. 39–46.

Santer, BD, Taylor, KE, Wigley, TML, Johns, TC, Jones, PD, Karoly, DJ, Mitchell, JFB, Oort, AH, Penner, JE, Ramaswamy, V, Schwarzkopf, MD, Stouffer, RJ & Tett, S 1996, 'Human effect on global climate?', *Nature*, vol. 384, pp. 524.

Stevens, B 2015, 'Rethinking the lower bound on aerosol radiative forcing', *Journal of Climate*, vol. 28, pp. 4794–4819.

Stocker, TF, Quin, D, Plattner, GK, Tignor, M, Allen, SK, Boschung, J, Nauels, A, Xia, Y, Bex, Midgely, PM (eds.) 2013, *Climate Change 2013: The Physical Science Basis, Contribution of Working Group I to the Fifth Assessment Report of the Intergovernmental Panel on Climate Change*, Cambridge University Press, Cambridge, United Kingdom and New York, NY, USA.

Watts, AW, Jones, EM, Nielsen-Gammon, JW & Christy, JR 2015, 'Comparison of Temperature Trends Using an Unperturbed Subset of The U.S. Historical Climatology Network', American Geophysical Union Fall Meeting, San Francisco, California, 14 December.

19: The Contribution of Carbon Dioxide to Global Warming

Arrhenius, S 1896, 'On the Influence of Carbonic Acid in the Air upon the Temperature of the Ground', *Philosophical Magazine and Journal of Science*, vol. 41, no. 5, pp. 237-276.

Barrett, J 1985, 'Paper on Spectra of Carbon Dioxide', *Villach Conference*, Austria, October 6-19.

Barrett, J 2005, 'Greenhouse molecules, their spectra and function in the atmosphere', *Energy & Environment*, vol. 16, no. 6.

Callendar, GS 1938, 'The artificial production of carbon dioxide and its influence on temperature', *Quarterly Journal of the Royal Meteorological Society*, vol. 64, pp. 223-240.

Callendar, GS 1941, 'Infra-Red Absorption by Carbon Dioxide, with Special Reference to Atmospheric Radiation', *Quarterly Journal of the Royal Meteorological Society*, vol. 67, pp. 263-275.

John, C, Herman, B, Pielke Sr, R, Klotzbach, P, McNider, RT, Hilo, JJ, Spencer, RW, Chase, T & Douglass, D 2010, 'What Do Observational Datasets Say about Modelled Tropospheric Temperature Trends since 1979?' *Remote Sensing*, vol. 2, iss. 9, viewed 29 March 2017, http://www.nsstc.uah.edu/users/john.christy/christy/2010_Christy_RS_Tropics.pdf.

DiChristopher, T 2017, 'EPA chief Scott Pruitt says carbon dioxide is not a primary contributor to global warming', *CNBC*, 9 March, viewed 28 March 2017, http://www.cnbc.com/2017/03/09/epa-chief-scott-pruitt.html.

Douglass, DH & Christy, JR 2009, 'Limits on CO2 climate forcing from recent temperature data of Earth', *Energy & Environment*, vol. 20, pp. 178-189.

Ekholm, N 1901, 'On The Variations of The Climate of the Geological and Historical Past and their Causes', Quarterly Journal of the Royal Meteorological Society, vol. 27, no. 117, pp. 1-62.

Fourier, J 1824, 'Remarques generales sur les Temperatures du globe terrestre et des espaces planetaires,' *Annales de Chimie et de Physique*, tome XXVII, pp. 136-167.

Fourier, J 1827, 'Mémoire sur Les Temperatures du Globe Terrestre et Des Espaces Planetaires', *Mémoires d l'Académie Royale des Sciences de l'Institute de France* tome VII, pp. 570-604.

Houghton, JT, Jenkins GJ & Ephraums JJ (eds) 2001, 'Climate Change: The IPCC scientific assessment', Cambridge University Press, Cambridge.

IPCC 2007, 'Summary for Policymakers', *Climate Change 2007: The Physical Science Basis*, Contribution of Working Group I to the Fourth Assessment Report of the Intergovernmental Panel on Climate Change, Cambridge University Press, Cambridge, United Kingdom and New York, NY, USA, viewed 28 March 2017, https://www.ipcc.ch/pdf/assessment-report/ar4/wg1/ar4-wg1-spm.pdf.

Kawamura, Y 2016, 'Measurement system for the radiative forcing of greenhouse gases in a laboratory scale', *Review of Scientific Instruments*, vol. 87, no. 1.

Kuhn, HG & Lewis, EL 1967, 'Self Broadening and f-Values in the Spectrum of Neon', *Proceedings of the Royal Society*, 25 July, vol 299, iss. 1459.

Lack, M 2013, *The Denial of Science*, AuthorHouse, Bloomington.

Laubereau, A & Iglev H 2013, 'On the direct impact of the CO_2 concentration rise to the global warming', *EPL*, vol. 104, no. 2.

Lewis, E, Rebbeck, MM & Vaughan, JM 1968, 'Resonance broadening in the spectrum of potassium', *Journal of Physics B: Atomic and Molecular Physics*, vol. 4, no. 5.

Lightfoot, HD & Mamer, OA 2014, 'Calculation of Atmospheric Radiative Forcing (Warming Effect) Of Carbon Dioxide at any Concentration Energy & Environment', *Energy & Environment*, vol. 25, no. 8, pp. 1439-1454.

Lindzen, RS 1981, 'Turbulence and Stress Owing to Gravity Wave and Tidal Breakdown', *Journal of Geophysical Research*, vol. 86, pp. 9707-9714.

Lindzen, R, Choi, M-D & Hou, AY 2001, 'Does the Earth Have an Adaptive Infrared Iris?' *Bulletin of the American Meteorological Society*, vol. 82, no. 3, pp. 417-432.

Lorentz, HA 1897, 'Ueber den Einfluss magnetischer Kräfte auf die Emission des Lichtes', *Annalen der Physik*, vol. 299, pp. 278-284.

Malvern, AR, Nicol, JL & Stacey, DN 1975, 'The effects of excitation mechanisms of spectral profiles in helium', *Journal of Physics B: Atomic and Molecular Physics*, vol. 7, no. 18, pp. 518-521.

United States Environment Protection Agency (EPA) n.d., *Causes of Climate Change*, viewed 28 March 2017, https://www.epa.gov/climate-change-science/causes-climate-change.

United States Environment Protection Agency (EPA) 2016, *Climate Change Indicators: Climate forcing*, viewed 21 March 2017, https://www.epa.gov/climate-indicators/climate-change-indicators-climate-forcing.

Wikipedia n.d., Electromagnetic spectrum, viewed 28 March 2017, https://en.wikipedia.org/wiki/Electromagnetic_spectrum.

21: The Geological Context of Natural Climate Change

Agnihotri, R & Dutta, K 2003, 'Centennial scale variations in rainfall (Indian, east equatorial and Chinese monsoons): Manifestations of solar variability', *Current Science*, vol. 85, pp. 459–463.

Alexander, WJR 2005, 'Linkages between Solar Activity and climatic Responses', *Energy and Environment*, vol. 16, pp. 239–253.

Archibald, D 2006, 'Solar Cycles 24 &25 and predicted climate response', *Energy and Environment*, vol. 17, pp. 29–38.

Archibald, D 2007, 'Climate outlook to 2030', *Energy and Environment*, vol. 18, pp. 615–619.

Archibald, D 2009, 'NASA now saying that a Dalton Minimum repeat is possible', *Solar Cycle 25*, 29 July, viewed 10 March 2017, http://sc25.com/index.php?id=51.

Archibald, D 2010, 'Solar Cycle 24 Update', *Solar Cycle 25*, viewed 10 March 2017, https://wattsupwiththat.com/2010/02/02/solar-cycle-24-update/.

Augustin, L, Barbante, C, Barnes, PR et al. 2004, 'Eight glacial cycles from an Antarctic ice core', *Nature*, vol. 429, pp. 623–628.

Avery, DT & Singer SF 2008, *Unstoppable Global Warming: Every 1,500 Years*, 2nd edn, Rowman & Littlefield Publishers.

Berger, A & Loutre, MF 1991, 'Insolation values for the climate of the last 10 million years', *Quaternary Science Reviews*, vol. 10, pp. 297–317.

Bond, G, Kromer, B, Beer, J, Muscheler, R, Evans, MN, Showers, W, Hoffmann, S, Lotti-Bond, R, Hajdas, I & Bonani, G 2001, 'Persistent solar influence on North Atlantic climate during the Holocene', *Science*, vol. 294, pp. 2130–2136.

Brandsma, T, Konnen, GP & Wessels, HRA 2003, 'UHI effects', *International Journal of Climatology*, vol. 23, pp. 829–845.

Brauer, A, Haug, GH, Dulski, P, Sigman, DM & Negendank, JFW 2008 'An abrupt wind shift in western Europe at the onset of the Younger Dryas cold period', *Nature Geoscience*, vol. 1, no. 8, p. 520.

Broecker, WS, 2006 'Abrupt climate change revisited', *Global and Planetary Change*, doi:10.1016/j.gloplacha.2006.06.01.

Butler, CJ & Johnston, DJ 1996, 'A provisional long mean air temperature series for Armagh Observatory', *Irish Astronomical Journal*, vol. 21, pp. 251–273.

Carter, RM, De Freitas, CR, Goklany, IM, Holland, D & Lindzen, RS 2007, 'Climate change. Climate science and the Stern Review', *World Economics*, vol. 8, pp. 161–182.

Cini Gastagnoli, G, Taricco, C & Alessio, S 2005, 'Isotopic record in marine shallow-water core: imprint of solar centennial cycles in the past 2 millennia', *Advances in Space Research*, vol. 35, pp. 504–508.

Clilverd, MA, Clarke, E, Ulrich, T, Rishbeth, H & Jarvis, MJ 2006, 'Predicting Solar Cycle 24 and beyond', *Space Weather*, vol. 4, iss. 9, pp. 1–7.

Dansgaard, W, Johnsen, SJ, Clausen, HB & Langway Jr, CC 1971, 'Climate record revealed by the Camp Century ice core', in KK Turekian (ed.), *Late Cenozoic Glacial Ages Symposium*, Yale, UP, pp. 37–56.

Dansgaard, W, Johnsen, SJ, Moller, J & Langway Jr, CC 1969, 'One thousand centuries of climate record from Camp Century on the Greenland Ice Sheet', *Science*, vol. 166, pp. 377–381.

Davis, BAS, Brewer, S, Stevenson, AC & Guiot, J 2003, 'The temperature of Europe during the Holocene reconstructed from the pollen data', *Quaternary Science Reviews*, vol. 22, pp. 1701–1716.

Emiliani, C 1955, 'Pleistocene temperatures', *Journal of Geology*, vol. 63, pp. 538–578.

Fall, S, Niyogi D, Gluhovsky, A, Pielke Sr, RA, Kalnay, E & Rochon, G 2009, 'Impacts of land use and land cover on temperature trends over the continental United States: Assessment using North American Regional Reanalysis', *International Journal of Climatology*, vol. 30, pp. 1980–1993, doi:10.1002/joc.1996.

Friis-Christensen, E & Lassen, K 1991, 'Length of the Solar Cycle: An Indicator of Solar Activity Closely Associated with Climate', *Science*, vol. 254, pp. 698–700.

Graham, S 2000, 'On the shoulders of Giants: Milutin Milankovitch (1879–1958)', March 24, NASA Earth Observatory, viewed 10 March 2017, http://earth observatory.nasa.gov/Features/Milankovitch/.

Hays, JD, Imbrie, J & Shackleton, NJ 1976, 'Variations in the Earth's Orbit', *Science*, vol. 194, pp. 1121–1132.

Hingane, LS 1996, 'Is a signature of socio-economic impact written on the climate?' *Climatic Change*, vol. 32, pp. 91–102.

Idso, C 2010, 'Medieval Warm Period Project', viewed 10 March 2017, http://www.co2science.org/data/mwp/mwpp.php.

Kaufman, DS, Ager, TA, Anderson, NJ, Anderson, PM, Andrews, JT, Bartlein, PJ, Brubaker, LB, Coats, LL, Cwynar, LC, Duvall, ML, Dyke, AS, Edwards, ME, Eisner, WR, Gajewski, K, Geirsdóttir, A, Hu, FS, Jennings, AE, Kaplan, MR, Kerwin, MW, Lozhkin, AV, MacDonald, GM, Miller, GH, Mock, CJ, Oswald, WW, Otto-Bliesner, BL, Porinchu, DF, Rühland, K, Smol, JP, Steig, EJ & Wolfey, BB 2004, 'Holocene thermal maximum in the western Arctic (0–180°W)', *Quaternary Science Reviews*, vol. 23, pp. 529–560.

Keeling, CD & Whorf, TP 2000, 'The 1,800-year oceanic tidal cycle: a possible cause of rapid climate change', *Proceedings of the US National Academy of Sciences*, vol. 97, pp. 3814–3819.

Lambeck, K & Cazenave, A 1976, 'Long-term variations in the length of day and climatic change', *Geophysical Journal of the Royal Astronomical Society*, vol. 46, pp. 555–573.

Lassen, K 2009, 'Long-term variations in solar activity and their apparent effect on the Earth's climate', viewed 10 March 2017, http://www.tmgnow.com/repository/solar/lassen1.html.

Lie, O & Paasche, O 2006, 'How extreme was northern hemisphere seasonality during the Younger Dryas?', *Quaternary Science Review*, vol. 24, pp. 1159–1182.

Lui, J, Wang, B et al. 2009, 'Centennial scale variations of the global monsoon precipitation in the last millennium: results from ECHO-G model', *Journal of Climate*, vol. 22, pp. 2356–2371.

Marshall Space Flight Center, NASA 2010, 'Solar Cycle Prediction', viewed 10 March, 2017, https://solarscience.msfc.nasa.gov/predict.shtml.

McKitrick, R & Michaels, P 2004, 'A test of corrections for extraneous signals in gridded surface temperature data', *Climate Research*, vol. 26, pp. 159–173.

McKitrick, RR & Michaels, PJ 2007, 'Quantifying the influence of anthropogenic surface processes and inhomogeneities on gridded global climate data', *Journal of Geophysical Research*, vol. 112.

Meehl, GA, Arblaster, JM, Matthes, K, Sassi, F & van Loon, H 2009, 'Amplifying the Pacific Climate System Response to a Small 11-Year Solar Cycle Forcing', *Science*, vol. 325, pp. 1114–1118.

Moros, M et al. 2009, 'Holocene climate variability in the Southern Ocean recorded in a deep-sea sediment core off South Australia', *Quaternary Science Reviews*, vol. 28, pp. 1932–1940.

Mudelsee, M 2001, 'The phase relations among atmospheric CO_2 content, temperature and global ice volume over the past 420 ka', *Quaternary Science Reviews*, vol. 20, pp. 583–589.

Neff, U, Burns, SJ, Mangini, A, Mudelsee, M, Fleitmann, D 2001, 'Strong coherence between solar variability and monsoon in Oman between 9 and 6 kyr ago', *Nature*, vol. 411, pp. 290–293.

Pielke Sr, RA 2002, 'Overlooked issues in the US National Climate and IPCC Assessment: an editorial essay', *Climate Change*, vol. 52, pp. 1–11.

Pielke Sr, RA Davey, CA, Niyogi, D et al. 2007, 'Unresolved issues with the assessment of multi-decadal global land surface temperature trends', *Journal of Geophysical Research*, vol. 112.

Rahmstorf, S 2003, 'Timing of abrupt climate change: A precise clock', *Geophysical Research Letters*, vol. 30, no. 10, p. 1510.

Sclater, JG, Jaupart, C & Galson, D 1980, 'The heat flow through oceanic and continental crust and the heat loss of the Earth', *Reviews of Geophysics and Space Physics* 18, 269–311.

Shackleton, NJ & Opdyke, ND 1973, 'Oxygen isotope and palaeomagnetic stratigraphy of equatorial Pacific core V28-238: oxygen isotope temperatures and ice volumes on a 10^5 and 10^6 year scale', *Quaternary Research*, vol. 3, pp. 39–55.

Skilbeck, CG, Rolph, TC, Hill, N, Woods, J & Wilkens, RH 2005, 'Holocene millennial/centennial-scale multi-proxy cyclcity in temperate eastern Australian estuary sediments', *Journal of Quaternary Science*, vol. 20, pp. 327–347.

Soon, WH 2005, 'Variable solar irradiance as a plausible agent for multi-decadal variations in the Arctic-wide surface air temperature record of the past 130 years', *Geophysical Research Letters*, vol. 32.

Soon, W 2009, 'Solar Arctic-mediated climate variation on multi-decadal to centennial timescales: Empirical evidence, mechanistic explanation, and testable consequences', *Physical Geography*, vol. 30, pp. 144–184.

Soon, W & Baliunas, S 2003, 'Proxy climatic and environmental changes of the past 1000 years', *Climate Research*, vol. 23, pp. 80–110.

Soon, WH, Posmentier, ES & Baliunas, SL 1996, 'Inference of solar irradiance variability from terrestrial temperature changes, 1880–1993: An astrophysical application of the sun–climate connection', *The Astrophysical Journal*, vol. 472, pp. 891–902.

Steffensen, JP, Andersen, KK, Bigler, M, Clausen, HB, Dahl-Jensen, D et al. 2008, 'High-resolution Greenland ice core data show abrupt climate change happens in a few years', *Science Express*, vol. 321, iss. 5889, pp. 680–684.

Steyaert, LT & Knox, RG 2008, 'Reconstructed historical land cover and biophysical parameters for studies of land-atmosphere interactions within the eastern United States', *Journal of Geophysical Research*, vol. 113.

Svensmark, H, Olaf, J, Pedersen, P, Marsh, N, Enghoff, M & Uggerhøj, U 2007, 'Experimental Evidence for the role of ions in particle nucleation under atmospheric conditions', *Proceedings of the Royal Society of London*, A463, pp. 385–396.

Taylor, KC, Lamorey, GW, Doyle, GA, Alley, RB, Grootes, PM, Mayewski, PA, White, JWC & Barlow, LK 1993, 'The flickering switch of late Pleistocene climate change', *Nature*, vol. 361, pp. 432–435.

Watanabe, O, Jouzel, J, Johnsen, S, Parrenin, F, Shoji, H & Yoshida, N 2003, 'Homogenous climate variability across East Antarctic over the past three glacial cycles', *Nature*, vol. 422, pp. 509–512.

Watts, A 2010, 'Solar geomagnetic index reaches unprecedented low – only 'zero' could be lower – in a month when sunspots became more active', viewed 10 March 2017, https://wattsupwiththat.com/2010/01/07/suns-magnetic-index-reaches-unprecedent-low-only-zero-could-be-lower-in-a-month-when-sunspots-became-more-active/.

Willard, DA, Bernhardt, CE, Korejwo, DA & Meyers, SR 2005, 'Impact of millennial-scale Holocene climate variability on eastern North American terrestrial ecosystems: pollen-based climate reconstruction', *Global and Planetary Change*, vol. 47, pp. 17–35.

Williams, PW, King, DNT, Zhao, J-X & Collerson, KD 2004, 'Speleothem master chronologies: Combined Holocene ^{18}O and ^{13}C records from the North Island of New Zealand and their paleoenvironmental interpretation', *The Holocene*, vol. 14, pp. 194–208.

Yasuda, I 2009, 'The 18.6-year period moon-tidal cycle in Pacific Decadal Oscillation reconstruction from tree-rings in western North America', *Geophysical Research Letters*, vol. 36, L05605, doi:10.1029/2008GL036880.

CPSIA information can be obtained
at www.ICGtesting.com
Printed in the USA
LVOW06s1519200917
549413LV00006B/39/P

9 780909 536039

CROW'S RANGE

Crow's Range

An Environmental History of the Sierra Nevada

David Beesley

UNIVERSITY OF NEVADA PRESS ▲▲ RENO & LAS VEGAS

University of Nevada Press, Reno, Nevada 89557 USA
Copyright © 2004 by University of Nevada Press
All rights reserved
Manufactured in the United States of America
Design by Carrie House
Library of Congress Cataloging-in-Publication Data
Beesley, David, 1938–
Crow's range : an environmental history of the Sierra
Nevada / David Beesley. — 1st ed.
p. cm.
Includes bibliographical references and index.
ISBN 0-87417-562-3 (hardcover : alk. paper)
1. Human ecology—Sierra Nevada Region (Calif. and
Nev.) 2. Conservation of natural resources—Sierra
Nevada Region (Calif. and Nev.) 3. Sierra Nevada
Region (Calif. and Nev.)—Environmental conditions.
I. Title.
GF504.S54B44   2001
333.72'09794'4—dc22   2004006610
The paper used in this book meets the requirements
of American National Standard for Information Sci-
ences—Permanence of Paper for Printed Library Materials, ANSI z.48—1984. Binding materials were selected
for strength and durability.
FIRST PRINTING
13 12 11 10 09 08 07 06 05 04
5 4 3 2 1